Study Guide

for

Monroe, Wicander, and Hazlett's

Physical Geology
Exploring the Earth

Sixth Edition

Kathleen Devaney
El Paso Community College

Australia • Brazil • Canada • Mexico • Singapore • Spain • United Kingdom • United States

Printed in the United States of America

1 2 3 4 5 6 7 10 09 08 07 06

Printer: Thomson/West

0-495-01150-9

Cover image: Jim Wark, Airphoto

Thomson Higher Education
10 Davis Drive
Belmont, CA 94002-3098
USA

For more information about our products, contact us at:
Thomson Learning Academic Resource Center
1-800-423-0563

For permission to use material from this text or product, submit a request online at **http://www.thomsonrights.com.**
Any additional questions about permissions can be submitted by email to **thomsonrights@thomson.com.**

Table of Contents

Preface

Preface written by Christopher Haley
VA Wesleyan College

The purpose of this study guide is to assist the college student in the introductory level geology class with the sometimes daunting task of assimilating the material presented in the textbook. It is designed both to act as a guide while reading each chapter and to test you on the material after the chapter has been read and studied. To maximize its use read this preface before starting the first chapter.

STRATEGIES FOR SUCCESS: A SPECIAL INTRODUCTION FOR FIRST YEAR COLLEGE STUDENTS.

The first year of college is always an exciting experience, but often a frustrating one as well. For many, it is the first time you will make many decisions that were made for you by others in the past. You now have to decide when to do homework, when to watch TV, when to go to parties, whether or not to go to class. While the freedom is refreshing, the distractions are many and many college freshmen find themselves not doing as well academically as they thought they would. It is important to be aware that the level of achievement expected for an A or a B grade in college is generally *much higher* than any previously encountered. Listed below are some suggestions on avoiding the pitfalls that many students fall into their first year.

1. GO TO CLASS, EVERY CLASS. In secondary school this was not a decision normally left to you. Although, there are no truant officers to make you attend class in college, I would suggest that any perception that you can get away with skipping class is an illusion. Skipping classes is the single biggest mistake that college students make their first year, and continue to make in subsequent years if they fail to recognize it as a problem. I would submit that skipping classes is a drug just as surely as alcohol or illegal narcotics. It is addicting and destructive, yet can be so easily rationalized in the mind of the individual that he or she is positive that the reasons for a bad grade in the class are everything *but* the failure to attend.

Some college instructors will deduct from your grade if you skip too many classes and they say so at the beginning of the semester. By and large, college administrations will back up an instructor's decision to deduct from a grade for this reason, so it is useless to appeal. On the other hand, many instructors do not require class attendance. These instructors feel that the grade achieved will probably reflect this lack of attendance and that a stated requirement is not necessary. Either way, in the "real world" there are dire consequences for skipping work. So, too, in college.

2. LEARN TO TAKE GOOD NOTES. There are several reasons for this. First and most importantly, no instructor has exactly the same vision of what is important as the authors of your textbook. Almost all instructors at some point present material that is not emphasized in the book, and for this your notes will be crucial. Second, your instructor may explain concepts also presented in the book in a way quite different from the way the book presents them. The more ways that you are presented material, the better the chances of you grasping the important concepts. Third, taking notes keeps you "in the ball game". In large classes, especially, where student participation can easily be avoided it is easy to start to notice things like chalk dust on the instructor's clothing, the student constantly sniffing to your left, the clown behind you making wisecracks. Concentration is difficult, but taking notes can help you focus.

Good and useful notes are aided greatly by sitting toward the front of the classroom. You can see and hear better and there are fewer distractions. This is particularly true in large lecture halls. Another good habit is to copy your notes over at the end of class. This makes them neater and gives you a chance to translate the "short hand" symbols in your notes that "you were sure you would remember what they meant". Some find that using multiple colors in their notes helps to separate different themes or to emphasize points that your instructor indicated were particularly important.

3. SHOW UP IN CLASS PREPARED. Generally this means reading the assignment before it is discussed in lecture. The text in your book is very concise. You will rarely, if ever, have to read more than one chapter before a

lecture. For most of the chapters in your book, this can be achieved in a couple of hours. Reading the textbook and using this study guide are a good way to spend that two to three hours. Having been introduced to vocabulary before a lecture makes the lecture clearer and more interesting. Study the diagrams and figures in your chapter before class as well. Geology is a very visual subject and the diagrams in your textbook are indeed, "worth a thousand words". It has long been a basic tenet of college that for every hour in the classroom you should plan on spending two to three hours studying on your own.

4. ASK QUESTIONS AND UTILIZE YOUR INSTRUCTOR'S OFFICE HOURS. There is no point in beating your head against the wall if there is some information that you just can't seem to assimilate. Many students have one major mistaken perception about asking questions: that the instructor and other students in the class will think the question (and therefore the student) is dumb. Forget about it. If you are lost, there is a good chance that you are not alone. Think about this, when was the last time that *you* thought that a fellow student was an idiot for speaking up and asking a question?

Nevertheless, if asking questions in the public forum of 100 or more students is not your style then use the instructor's office hours. These one-on-one encounters may pleasantly surprise you. For instance, you may find out that your instructor is a human being with concern and compassion for the student (it's been known to happen). You may find out that you are having trouble with material because you don't understand something else that came before, thereby avoiding a disaster. You actually may get to know the instructor which, somehow, often seems to help. Not enough students take advantage of office hours. Second only to your own mind, the best resources you have at college are the minds of your instructors. Don't fail to take advantage of that.

TIPS ON STUDYING FOR EXAMS

1. DON'T WAIT UNTIL THE LAST MINUTE. You need to leave ample time to get help from the instructor (see #4 above) if you suddenly discover that there is something that you don't understand. Waiting until the last minute inevitably leads to the "all nighter". Staying up all night before a test usually has the same effect on your performance that it would have on a Marathon runner's performance. Rest the night before the event is essential.

2. STUDY IN A GROUP, BUT WITH THESE GUIDELINES: Do not depend on the group to get you through the exam. They won't be there for you when you take the exam. Make sure your group stays focused on the task at hand. Discussing sports, fashion, relationships, the food at the dining hall, or which fraternity or sorority you are thinking about rushing can wait. If your study mates insist on talking about anything except geology, walk away. Leave some time to study alone after your group meets. Groups tend to decide what to go over by committee and this may not be what you need to work on most.

3. USE OLD EXAMS IF THEY ARE AVAILABLE. These might be useful if your instructor has put them on file (not that many do). However, you should only use them as a guide for the type of material that was emphasized and the types of questions asked. DO NOT memorize the questions and hope that they will be the same on your exam. If an instructor gives the same test year in and year out, you can bet that the test is not on file.

4. USE THIS STUDY GUIDE. Treat the questions in this study guide and those that follow each chapter in your textbook as sample tests. Some hints on how to maximize the use of this guide are given below.

USE OF THIS STUDY GUIDE

This study guide is meant to be a supplement to your textbook and notes. You should be under no delusion that use of this guide is in any way replacement for attending class and reading your text. That said, however, you should find it a very helpful guide to the information in your textbook. Each chapter is divided into six to nine parts. These include: 1) Chapter Objectives, 2) Useful Analogies (not in all chapters), 3) Key Terms, 4) Chapter Concept Questions, 5) Complete the Table (not in all chapters), 6) Completion Questions, 7) Multiple Choice, 8) True or False, 9) and Drawings and Figures (in most, but not all chapters)

CHAPTER OBJECTIVES

This is a list of goals that you should set for yourself when attacking a chapter. To maximize the use of this section you should read it over once *before* you start the chapter, and then keep it in front of you while you read the chapter. When you finish reading a section of the chapter, read the objectives that fit the information covered in that section. Satisfy yourself that you can fulfill that objective or at least that you understand what that objective entails. If you do not feel confident that you have achieved the objectives for some sections, be sure to reread those sections.

USEFUL ANALOGIES

Some concepts are just plain difficult to understand. Although the authors of your textbook and your instructor have worked hard at explaining these concepts clearly, sometimes they just don't stick. This section attempts to explain some of the more difficult concepts in something other than geologic terms. This use of analogy may prove useful for some students.

KEY TERMS

This is a list of geologic terms that are introduced in each chapter. Half the battle in conquering the subject of geology is mastering the terms. Not only should you know the definition of each term but you should also be able to use the terms. You should also know why each is significant. They are listed approximately in the order that they are discussed in the text. It is helpful to keep this list in front of you as you read the chapter, highlighting the terms as they are encountered in the text. The list is also useful to use with your notes. It is not uncommon for students to misspell words when taking notes, even to the point where they may be hard to recognize. At the very least this list provides proper spelling! When studying for exams you can use this list as a list of terms to define.

CHAPTER CONCEPT QUESTIONS

These are discussion questions that require some thought and organization to answer. Since the wording of each answer is a highly individual matter, there was no attempt to give a specific answer in the back of the book. Instead textbook page numbers are given in the Answer Appendix where information to help you answer the question may be found. You will find that many of these mirror the chapter objectives.

COMPLETE THE TABLE

These are included in chapters where some sort of classification is encountered. They are generally self-explanatory. Filling in the required blanks may prove helpful in learning these classification systems. Like the Chapter Concept Questions, page numbers or references to similar tables in the textbook are given in lieu of exact answers.

CHAPTER COMPLETION, MULTIPLE CHOICE, AND TRUE OR FALSE QUESTIONS.

These are sample questions in three formats that are commonly encountered on exams. The answers are given in the back of this manual. If your instructor does not tend to ask all of these types of questions, you should still look at them because many of them can be reworked into another type of question. For instance, completion questions can easily be made into multiple choice questions and vice versa. True-false questions commonly appear as multiple choice questions that begin with, "Which of the following is not true about…"

DRAWINGS AND FIGURES

Geology is a very visual subject. This fact is made evident by the lavish use of diagrams and photographs in your textbook. Undoubtedly your lecturer makes use of slides, PowerPoint, CD-ROMs, videos, or overheads. Some instructors incorporate illustrations into their tests or ask questions in which you are asked to sketch. To give you practice answering these types of questions most chapters include a section of this type.

OTHER RESOURCES

In addition to your textbook, professor and notes there may be other resources available to you at your school that you should be aware of. For example, does your school have a geology museum? If so, twenty minutes walking around it may prove one of the most productive ways to add to your knowledge or clarify information presented in class. As previously stated, geology is a very visual field. If a picture is worth a thousand words, then actual specimens are worth ten thousand.

If you are required to write out-of-class essays or papers, or if you are just interested in learning more about this field, some of the following suggestions might prove useful.

PHYSICAL GEOLOGY NOW™ WEB RESOURCE

The publishers of your textbook sponsor a website called PHYSICAL GEOLOGY NOW™. Although not available at press time, it will be ready by the time you are reading this. The concept behind the PGN promises to be innovative and extremely useful. The website promises to guide you through personalized learning plans, provide you with simulations and animations of key concepts, provide flashcards, and give practice exams. As you work through the text, you will see notes that direct you to dynamic, media enhanced activities and tutorials in PGN.

PERIODICALS

There is a plethora of professional journals in geology for virtually every discipline within the science. Many of these will prove too esoteric or complex for the average introductory student, but if you would like to have a look, your school library probably subscribes to several. Rather than list them here (there are way too many) I would suggest going and looking in your library. The library subject catalog or the reference librarian can help you find them.

There are other periodicals that are less directed to the professional geologist, that often carry geology-related articles. *National Geographic* often has articles pertinent to environmental issues, paleontology, and even geology. *Discover* is a good college level, general science publication that often has articles related to the earth sciences. *Scientific American* also is oriented toward science students. Its articles are often a bit more technical than the others listed but should not be beyond the comprehension of a college freshman. The National Parks Conservation Association publishes *National Parks;* a magazine devoted to informing the American public about one of our greatest legacies. Their articles are very informative and often contain a great deal of information on geologic features of our parks. This list is, of course, incomplete, so ask your librarian about other periodicals.

Most libraries have online databases that search a wide array of journals for specific topics. GEOREF and GEOBASE are two well-known examples specific to the geological literature. Info Trac also can lead to articles in more popular periodicals. Check with your librarian to find out which you have access to and learn to use them. Increasingly, journals are providing full-text articles for download through these databases. Unfortunately, as of this writing, the geological journals are not embracing this as much as some other disciplines but this will probably change over time.

VIDEOS AND TELEVISION

Cable television and public television often carry programs devoted to the earth. The Discovery Channel has built a reputation featuring quality science oriented shows and is carried by most cable operators. Similarly, public television often has specials or even whole series dedicated to the earth sciences. Nova, a weekly science program on most public television stations, offers very high quality entertainment of interest to the earth science student. New programs and cable channels are coming along all the time so the only advice I can give is to check your local listings.

You can also check your library for videos on earth science topics. Many libraries buy videos of some of the Nova programs. Some particularly good ones are Volcano!, In the Path of a Killer Volcano, Chasing El Niño, T. Rex Exposed, The Day the Earth Shook, Science Now (also available online), and Magnetic Storm. The excellent and popular series Earth Revealed is also available as a video series and is probably available in your library. It is also available online free from the Annenberg CPB website. There are many others, too numerous to mention so, once again, check your libraries, both college and public to see what they have.

ROADSIDE GEOLOGY GUIDES

The Mountain Press publishes guidebooks for those interested in learning about the geology of the places they drive through. They are written at a level that really requires no previous knowledge but are most useful for people that have had an introductory course (that is you, almost!). One is published per state (although some states such as Vermont and New Hampshire are grouped together). At this writing, not all states are covered by the series. According to the Mountain Press website the following states are currently available: Alaska, Arizona, California (northern and central), Colorado, Hawaii, Idaho, Indiana, Maine, Massachusetts, Montana, Nebraska, New Mexico, New York, Oregon, Pennsylvania, South Dakota, Texas, Utah, Vermont and New Hampshire, Virginia, Washington, Wisconsin and Wyoming. They also have one on the Yellowstone Country.

Mountain Press is also publishing a Geology Underfoot series for those that want to get off the highway. Currently Central Nevada, Death Valley and Owens Valley, Illinois, and Southern California are covered. Think how impressed your instructor will be if you write a paper on the local geology and actually visit various places in your area! New states are published every year. To find out if a Roadside Geology Guide is available your state, you can visit the Mountain Press Website at http://www.mtnpress.com.

THE INTERNET

The most rapidly growing source of information available to students in the last ten years is the Internet. Virtually every college has ready access to it in the library and an increasing number of schools are wiring college dormitories so that all students can access the "web" with minimum effort. Finding information has become extremely easy in the past few years with the advent of search engines. At the time of this writing Google is probably the most useful and popular among scientists. In this day and age it is hard to imagine that you have not had abundant experience using the Internet for fun and learning. If, somehow, you managed to miss the revolution, or just haven't used it enough to become really good at it, your library can help you there to.

The greatest problem with using the Internet these days is not in finding information about a particular subject, but in evaluating what information is good and what is unreliable. For example, any attempt to find information on fossils, earth history, or evolution is bound to result in a large number of pages created by "creation scientists". While this is not the place to debate the validity of this view of Earth, chances are that a term paper filled with such references will not impress the professor. Sites from government science organizations are reliable, up to date and informative. The Worldwide Web has become *the* vehicle by which government agencies are communicating what they do to the public. The following sites are great places to start:
The National Oceanic and Atmospheric Administration – http://www.noaa.gov
The U.S. Geological Survey – http://www.usgs.gov
The Environmental Protection Agency – http://www.epa.gov
The Federal Emergency Management Agency – http://www.fema.gov
The National Aeronautics and Space Administration – http://www.nasa.gov
The Bureau of Land Management – http://www.blm.gov
Sites associated with colleges and universities are often excellent. University of North Dakota, for example, operates Volcano World, (http://volcano.und.edu), a fun interactive and informative site where you can find information and photos of virtually every volcano in the world. There are many others and new ones are starting up all the time. The publisher of your textbook maintains a site (http://www.earthscience.brookscole.com/physgeo5e that has many links to geology related sites that have a direct bearing on material covered in your textbook. This is a selective list and you can trust the sites in it to be appropriate. Before using any site in a research paper, however, it is best to clear it with your instructor.

ACKNOWLEDGEMENTS

By Kathleen Devaney
El Paso Community College

This edition is an update of Christopher Haley's *Study Guide for Monroe and Wicander's Physical Geology, Fifth Edition*; with new questions added to those written by Mr. Haley and the book reorganized to match Monroe and Wicander's *Physical Geology, Sixth Edition.*

In updating this edition, many thanks go to Ms. Carol Benedict, editor at BrooksCole, and Mr. John Banks for all their help and patience.

Chapter 1

Understanding Earth: A Dynamic and Evolving Planet

CHAPTER OBJECTIVES

By the end of this chapter you should be able to:

1. Appreciate Earth as a system and describe several subsystems that contribute to the larger Earth system.
2. Describe several professional endeavors in which geologists are involved.
3. Discuss how some knowledge of geology and the earth can help any citizen as a consumer and a voter.
4. Describe several environmental issues facing humankind that are linked to geology.
5. Summarize the currently accepted theory for the origin of the universe, solar system, and Earth.
6. Name, compare, and contrast the different layers of Earth.
7. Compare and contrast the two types of crust.
8. Summarize the scientific method and differentiate between a hypothesis and a theory.
9. Appreciate the impact of Plate Tectonic Theory in understanding the earth.
10. Distinguish between the different types of plate boundaries on the basis of relative movement of the plates on either side.
11. Discuss the relationship among the three kinds of rocks and the various processes operating on and within Earth, which transfer materials from one kind of rock to another.
12. Explain the principle of uniformitarianism.

KEY TERMS

After reading and studying this chapter you should know the following terms:

system
geology
physical geology
historical geology
topography
sustainable development
greenhouse effect
Kyoto Protocol
Big Bang
gravity
electromagnetic force
strong nuclear force
weak nuclear force
solar nebula theory
solar nebula
planetesimal
terrestrial planets
Jovian planets
core
mantle
asthenosphere
lithosphere
crust
plates
convection cells
continental crust
oceanic crust
theory

scientific method
hypothesis
plate tectonic theory
divergent plate boundaries
converging plate boundaries
transform plate boundaries
rock
mineral
rock cycle
igneous rock
sedimentary rock
metamorphic rock
geologic time scale
uniformitarianism

CHAPTER CONCEPT QUESTIONS

1. List and define three subsystems of the larger Earth system and describe how they interact with one another.

Atmosphere-air Lithosphere-plates biosphere-global ecosystem

2. List and define 4 fields of specialization in geology.

Economic Geology-minerals/energy resources Paleontology-Fossils Planetary Geology-geology of planets

3. Describe three ways in which geology bears on everyday life.

Earthquakes, Volcanos Tsunamis, rocks gems jewels

? 4. How does *sustainable development* differ from past practices? Why is it important to the future of our society?

5. Describe several ways in which an ever-increasing world population has added stress to our environment. What role does geology play in addressing these issues?

housing, factories, tears down the environment / people in areas of danger / dwindling our non-renewable resources

6. Explain what the greenhouse effect is, its causes and its effects.

= retention of heat [illegible] trap heat reflected back from earths surface

7. According to the most widely accepted theory, how did the universe form?

Big bang. Big explosion then star dust formed & collected - Universe expanding

8. Where do the elements heavier than hydrogen and helium come from?

When the universe cooled after big bang → star explosions

9. According to the most widely accepted theory, how did our solar system form?

A Big bang of pure energy creating atoms → chemical make up of uni. changed

condensation then collapse of interstellar material

10. What are the main differences between the terrestrial planets and the Jovian planets?

Ter = earth like (rocks/metallic elements) Jov = small rock core w/ mostly gas

11. What is meant by the characterization of Earth as a differentiated planet and how did it get that way?

heat, impacts, gravity → created concentrated layers → made gases → plants etc

12. Describe how scientists define a question, carry out studies and draw conclusions.

Hypothesis, theory → scientific method

13. List the three types of plate boundaries and describe their differences in terms of relative plate motion.

Converge, Diverge Transform

14. What do geologists think is the mechanism by which plates move?

Mantle moving beneath the crust → convection currents

15. Using plate tectonics and the rock cycle as examples explain what is meant by the characterization of Earth as a "dynamic planet".

All interrelated Con → meta div → Igneous Trans → sed etc (6) cycle

16. Although plate tectonics is essentially a phenomenon of the solid Earth, it impacts other parts of our planet. Explain how plate tectonics affects the atmosphere, hydrosphere, and biosphere.

Gases in the air / changes the plates/crust → create mountains new living areas for animals

17. What is the general origin of the igneous, sedimentary, and metamorphic rocks?

Volcanos → divergent bs con bs batholiths etc Sed - near water Met = con

18. James Hutton's principle of Uniformitarianism is often stated as, "The present is the key to the past". Explain this statement.

- The things here in the present ex. the past
- Pangea = glaciers that are the same on dif cont today
- Mountains = explain plate tectonics

COMPLETE THE FOLLOWING TABLES:

Layer	% Earth	Avg. Density	Composition
Core	16	10-13g/cm3	Iron/nickel
Mantle	83%	3.3-5.7 g/cm3	Iron/magnesium
Crust	1%	2.7 g/cm3	silicon/aluminum

-small solid region -> larger liquid region

Crust Type	Thickness	Avg. Density	Composition
Continental			
Oceanic			

COMPLETION QUESTIONS

1. Mineralogy is the study of Minerals while petrology is the study of rocks.

2. Stratigraphy is the study of layered rocks/sediments.

3. Structural geology is the study of the rock deformation of the crust.

4. Paleontology is the study of fossils.

5. In the opinion of many scientists, an increase in carbon dioxide from burning fossil fuels over the past 200 years is a prime suspect in causing global warming.

6. The age of the universe is approximately 15 bil years.

7. Initially the universe was composed of 100% hydrogen and helium. Today it is still 98% those two elements.

8. The Earth formed 46 billion years ago.

9. Our solar system formed from a rotating disk-shaped cloud of gas and dust called a solar nebula

10. Most of the matter in this cloud condensed in the center to form the sun, but local eddies in the disk accumulated cold matter to form the early planetesimals

11. The scientific approach of investigation involves developing a tentative explanation or hypothesis. If this explanation is shown to be repeatedly correct, a theory is proposed to explain the phenomena.

12. The Plate Tectonic theory has provided a framework for interpreting the composition, structure, and internal processes of Earth on a global scale.

13. The mechanism for moving the plates across the Earth appears to be convection within the Mantle.

14. At divergent margins the plates are moving away from each other.

15. Sedimentary rocks originate by consolidation of rock fragments, precipitation of minerals from solution, or compaction of plant or animal remains.

16. Metamorphic rocks result from the alteration of pre-existing rocks by heat, pressure, and chemical activity of fluids.

17. The premise that present-day processes have operated throughout Earth's history is called ________

18. The Earth's outer layers are broken into pieces called plates that interact to form ________, ________ and mountains.

19. Most of the Earth is made of ________ that forms the layer called ________.

20. The Earth is a ________ and ________ planet with a long history.

21. Most scientists agree the greatest environmental problem today is ________ since there are over ________ billion people on the planet.

22. The early Earth was probably formed of a homogeneous mix of the elements ________, ________, ________, and ________ among others.

23. Our standard of living is directly dependent on consumption of ________ ________.

24. The Earth's layers are different ________ because they have different compositions, temperatures, and ________

25. Volcanoes are made of ________ rocks and are common when ocean and continent plates ________

MULTIPLE CHOICE

1. Which of the following is a pursuit in which professional geologists are employed?
   a. the search for fossil fuels
   b. the search for groundwater
   c. locating safe building sites in earthquake prone areas
   d. all of the above

2. Petrology is the study of
   a. petroleum.
   b. rocks.
   c. fossils.
   d. none of the above.

3. Paleontology is the study of
   a. ancient landforms.
   b. ancient rocks.
   c. ancient minerals.
   d. none of these.

4. Most scientists view the root cause of our increasing environmental problems to be
   a. the greenhouse effect.
   b. water pollution.
   c. overpopulation.
   d. diminishing energy resources.

5. The first atoms formed ________ the Big Bang.
   a. within one second of
   b. about 300,000 years after
   c. about 200 million years after
   d. about 15 billion years after

6. During formation of our solar system ___ % of the mass in the collapsing nebulae concentrated in the center to become the sun.
   a. <10
   b. 30
   c. 60
   d. 90

7. Which of the following is NOT a terrestrial planet?
   a. Venus
   b. Earth
   c. Mars
   d. Jupiter

8. The differentiation of the earth resulted in the formation of
   a. the layers of the earth.
   b. the atmosphere.
   c. the oceans.
   d. all of these

9. The continental crust is rich in
   a. silicon and magnesium.
   b. aluminum and magnesium.
   c. silicon and aluminum.
   d. iron and magnesium.

10. The mantle is rich in
    a. silicon and magnesium. – oceanic plate
    b. aluminum and magnesium.
    c. silicon and aluminum. – continental plate
    d. iron and magnesium.

11. The core of the Earth is rich in
    a. iron and magnesium.
    b. iron and nickel.
    c. silicon and aluminum.
    d. silicon and oxygen.

12. The asthenosphere is composed of – upper part of mantle under lithosphere → plastically + flows
    a. crust only.
    b. crust and the upper part of the mantle.
    c. crust and the lithosphere.
    d. part of the mantle only.

13. The lithosphere is composed of the
    a. crust only.
    b. crust and the entire mantle.
    c. crust and the asthenosphere.
    d. crust and the mantle above the asthenosphere. – only a little bit

14. Plates move toward each other at a _______ boundary.
    a. divergent
    b. convergent
    c. transform
    d. all of the above

15. Plates move away from each other at a _______ boundary.
    a. divergent
    b. convergent
    c. transform
    d. all of the above

16. Igneous rocks form from
    a. pre-existing rocks undergoing changes due to heat, pressure, and chemically active fluids.
    b. a cooling magma or lava.
    c. sediments compacted and cemented.
    d. all of these

17. __________ rocks were once some other kind of rock until subjected to a large increase in pressure and/or temperature.
    a. igneous
    b. sedimentary
    c. metamorphic
    d. all of the above

18. Absolute ages have been assigned to the geologic time scale since the discovery of:
    a. plate tectonic theory.
    b. the first fossils.
    c. radioactivity.
    d. uniformitarianism.

19. The universe is
    a. colder than expected.
    b. shrinking due to powerful gravitational forces.
    c. about 20 times older than the Earth.
    d. still expanding after the Big Bang.

20. Plate movements have
    a. created mountains like the Andes.
    b. formed volcanoes.
    c. altered the distribution of life on Earth.
    d. all of these answers

21. Theories in science are
    a. explanations of nature based on evidence.
    b. wild guesses based on philosophy.
    c. only useful to explain past behavior.
    d. all of these answers

22. The rock cycle
    a. describes how rocks form.
    b. shows that all igneous rocks become metamorphic eventually.
    c. only operates at convergent boundaries.
    d. answers A and C only

23. Increase in atmospheric carbon dioxide
    a. is due in part to the burning of fossil fuels.
    b. is partly a result of deforestation.
    c. will continue.
    d. all of these answers

24. When scientists call the Earth "dynamic," they mean
    a. it's spinning.
    b. it is constantly changing.
    c. life today is really fast paced.
    d. it formed from the solar nebula.

25. Convection
    a. systematically removes iron from the crust.
    b. is caused by earthquakes.
    c. moves the plates.
    d. breaks the asthenosphere into plates.

TRUE OR FALSE

T 1. Most of the energy resources now being consumed are located by geologists.

F 2. Dinosaurs and humans lived together on Pangaea.

F 3. Most of the material from the solar nebula, that formed our solar system, is found in the planets.

T 4. The terrestrial planets are closer to the sun than the Jovian planets.

T 5. The Earth is composed of layers of different thickness and composition.

F 6. The outer core is the thickest layer of the earth.

F 7. The asthenosphere rests above the lithosphere.

T 8. The oceanic crust is denser than the continental crust.

T 9. Continental crust is thicker than oceanic crust.

T 10. The core is denser than the mantle.

F 11. The asthenosphere includes the crust and the upper most part of the mantle.

F 12. A hypothesis can only be developed, but it cannot be tested.

T 13. Plates slide past one another at transform plate boundaries.

F 14. The hydrosphere and atmosphere of Earth are independent of the solid Earth and are unaffected by Plate tectonics.

F 15. Sedimentary rocks form from the melting of pre-existing rocks.

F 16. Metamorphic rocks cannot undergo weathering.

F 17. The principle of Uniformitarianism requires that present day geologic processes have always operated at a uniform rate.

F 18. The asthenosphere is the only liquid layer of the Earth.

T 19. The Earth's crust is mostly made of less than 20 kinds of minerals.

T 20. Plate tectonics drives the rock cycle.

F 21. Scientists are constantly collecting information and testing the power of theories to explain that information.

F 22. The Earth's interior heat drives convection cells that exert drag on the plates and slow them down.

T 23. Organic evolution recognizes that organisms living today are descendants of those alive in the past.

___24. Once an hypothesis achieves theory status it is inviolate.

T 25. A knowledge of geology will help you invest in land and understand environmental issues.

DRAWINGS AND FIGURES

1. Reproduce the rock cycle as illustrate in Figure 1-13. Be sure to show all possible paths and label rock types and processes. Your version doesn't need to be as fancy as that in the book, but it should be accurate.

2. In the figure to the right label the layers of the Earth. In the inset in the upper right label the lithosphere, asthenosphere, upper mantle and the two types of crust. Check your answer against Figure 1-9 in your textbook.

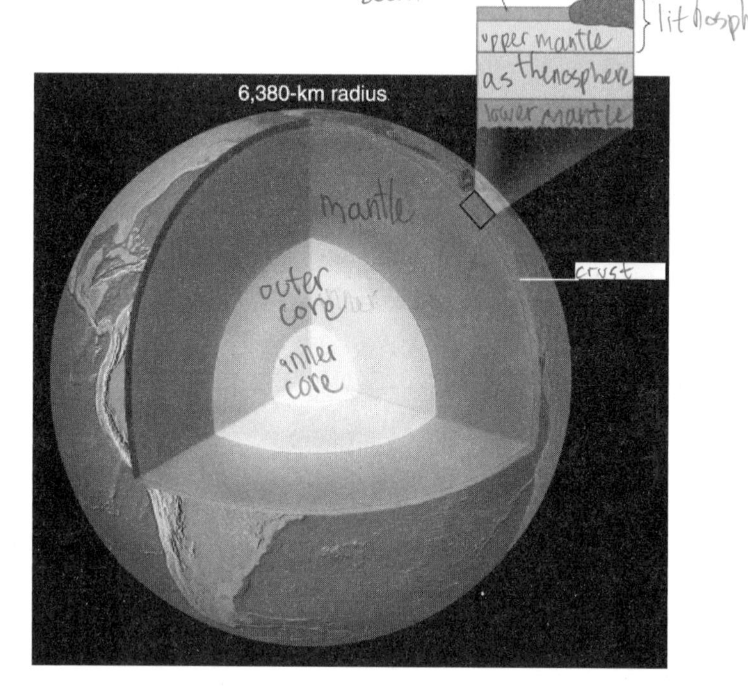

Chapter 2

Plate Tectonics: A Unifying Theory

CHAPTER OBJECTIVES

By the end of this chapter you should be able to:

1. List some of the early evidence that the continents of Africa and South America were once together, which provided the seeds of what was to become the continental drift hypothesis.
2. List the evidence that Wegner used to argue for his hypothesis of continental drift.
3. Discuss how the pioneering work in paleomagnetism of S. K. Runcorn supported the hypothesis of continental drift.
4. Summarize the theory of sea floor spreading as postulated by Harry Hess and list his evidence for it.
5. Summarize the paleomagnetic evidence for sea floor spreading contributed by Vine, Matthews and Morley.
6. Summarize the evidence in favor of sea floor spreading contributed by the Deep Sea Drilling Project.
7. Characterize the different types of plate boundaries recognized in plate tectonic theory.
8. Discuss several ways in which the past and present rate and direction of plate movement can be determined.
9. Summarize the alternative mechanisms for plate movement currently being discussed by geologists.
10. Rationalize the apparent relationship between plate tectonic boundaries and valuable metallic mineral deposits.

A USEFUL ANALOGY

Our Eatable Earth

You can build your own subduction zone to illustrate the accretion of a subduction complex to a continental margin. Find two tables that are not quite the same height. The two tables must be almost the same height so that the table tops overlap in thickness. In other words, as you push a sheet of cardboard from the shorter table toward the taller table, the cardboard will not be able to slide onto the top of the higher table but will not slide under the table top either; it will slide into the side of the table top. Set these two tables beside each other so that there is barely a gap between them.

Now, take a table cloth and put it on the shorter table so that it hangs between the tables. The shorter table is oceanic crust, the taller is continental crust, and the tiny gap between them is the trench.

Spread some peanut butter across the top of the oceanic crust getting plenty in the trench. This is sediment deposited just off the edge of the continent. Send some worthy victim, such as a younger sibling, below to pull the table cloth down through the trench. This is the oceanic crust moving beneath the continental crust. As this happens, the sediment (peanut butter) will not be able to go down the trench and will pile up in a highly contorted manner at the edge of the continent....instant subduction complex! You can vary the experiment by adding pie or cakes as microcontinents to raft into the continent.

KEY TERMS

After reading this chapter, you should be familiar with the following terms.

Gondwana
Glossopteris flora
continental drift
Pangaea
Laurasia
Mesosaurus
Lystrosaurus
Cynognathus
polar wandering
sea-floor spreading
thermal convection cells
Glomar Challenger
plate tectonics
plates
supercontinent cycle hypothesis
divergent plate boundary
convergent plate boundary
transform plate boundary
spreading ridges
subduction
Benioff Zone
oceanic-oceanic boundary
subduction complex
paleographic maps
back-arc basin
oceanic-continental boundary
continental-continental
mélange
ophiolite
transform faults
hot spots
mantle plumes
aseismic ridges
ridge push
slab pull

CHAPTER CONCEPT QUESTIONS

1. Summarize the evidence cited by Wegner, du Toit, and others as support for the hypothesis of continental drift. Include evidence suggested by map patterns, fossils, rock sequences, mountain ranges, and Paleozoic glaciation.

2. If so much evidence had accumulated in support of the original idea of continental drift, why did geologists reject this idea?

3. What are the alternative explanations for the variation in the orientation of magnetic minerals in rocks of different ages? Which of these is the best explanation and why?

4. How does the theory of sea floor spreading differ from continental drift? What new observations in the 1960s lead to the development of this theory?

5. What do the magnetic "stripes" across the sea floor indicate? How do they support the theory of sea floor spreading?

6. Compare the age of the oceanic crust to the age of the continental crust. How does this data support sea floor spreading?

7. What is meant by the assertion that Plate Tectonics is a unifying theory?

8. Explain the theory of the supercontinent cycle. Cite evidence supporting this theory.

9. Summarize succinctly but completely, the characteristics of divergent plate boundaries.

10. Enumerate the steps in the rifting of a continent and the opening of an ocean basin.

11. What geologic features can be used by geologists to identify sites of ancient rifting?

12. Compare and contrast the characteristics of ocean-ocean, ocean-continent, and continent-continent convergent margins.

13. What geologic features can be used by geologists to identify sites of ancient convergent plate boundaries?

14. What are ophiolites? What have they contributed to our knowledge of the oceanic crust?

15. How can the speed and direction of motion of a plate be determined?

16. Describe the alternative driving mechanisms for plate tectonics?

17. How are the locations of mineral deposits influenced by plate tectonics?

COMPLETION QUESTIONS

1. The abundant fossils of the ________ flora were used as evidence that the southern hemisphere continents were once together.

2. When the present continents were joined together, they formed the super continent of ________, with the northern hemisphere continents forming ________, and the southern hemisphere continents forming ________.

3. The father of continental drift is generally considered to be ________ thanks to his book, The Origin of Continents and Oceans.

4. When reconstructing Pangea, the best fit is achieved when joining the present day continental margins at a depth of ________ below sea level.

5. The orientation of glacial striations indicates that the late Paleozoic south pole was located in present-day ________.

6. The discovery of the great extent of the mid-ocean ridge systems prompted Harry Hess to propose the theory of ________.

7. The symmetric disposition of alternating positive and negative ________ about the ________ is one of the most convincing pieces of evidence for the theories of sea floor spreading and plate tectonics.

8. The oldest sea floor is less than ________ million years old and the oldest continental crust is ________ old.

9. The interactions between plates dictate the positions of continents, ________, and mountains and therefore affect Earth's ____________.

10. Sea floor spreading states that sea floor ________ at an oceanic ridge, and is ________ in a trench.

11. Plate tectonics states that plates are large slabs of ________, which float on semiplastic ________.

12. As a result of sea floor spreading, with increasing distance form a ridge crest, the sea floor becomes ________ in age and the sediments become ________ in thickness.

13. The main rock type produced at oceanic ridges and, therefore, making up the sea floor is ________.

14. When a continent undergoes extension, a large linear depression called a ________ forms. A good example of this occurs in Eastern Africa.

15. Subduction zones are characterized by dipping planes of earthquake foci called ________ zones.

16. As sea floor is subducted, it partially melts. The resulting magma rises to the surface to produce volcanic rocks that are ________ in composition.

17. A curved chain of islands called an ________ forms over subduction zones at ocean-ocean convergent margins.

18. The lithosphere on the landward side of a volcanic island arc may be stretched and thinned resulting in a ________.

19. As a plate is subducted, sediments are scraped off the descending plated at the trench and added to the overriding plate as a ________ complex.

20. Ancient convergent boundaries are sometimes recognized by the presence of highly deformed and metamorphosed submarine rocks known as ________.

21. During subduction, pieces of oceanic lithosphere called ________ may actually accrete onto continental crust.

22. At a transform boundary, two plates slide past one another, as shown by the ________ Fault in California.

23. Hot ________ are plumes of rising ________, coming from deep within the Earth.

24. The volcanoes on the Big Island of ________ have formed from a hot spot.

25. The driving mechanism for plate tectonics is not completely known, but ________ cell generated movement is a very popular theory.

MULTIPLE CHOICE

1. The Sumatran earthquake and tsunami of December 26, 2004
   a. was caused by the motion of tectonic plates.
   b. could happen again.
   c. could not be controlled since it was driven by forces deep within the Earth.
   d. all of these answers

2. All of the present day continents were at one time joined together to form
   a. Gondwanaland.
   b. Laurasia.
   c. Pangaea.
   d. none of the above

3. Glacial evidence from _________ supports the concept of Pangaea.
   a. the Northern Hemisphere
   b. the Southern Hemisphere
   c. Laurasia
   d. the Arctic

4. Glossopteris was
   a. a seed fern.
   b. an early mammal.
   c. type of dinosaur.
   d. none of the above

5. _____ proposed the theory of _____ in the early 1960s and this led very quickly to the theory of plate tectonics.
   a. Alfred Wegner / continental drift
   b. Alfred Wegner / sea floor spreading
   c. Harry Hess / continental drift
   d. Harry Hess / sea floor spreading

6. The best explanation for polar wandering curves is
   a. the continents were fixed as the north magnetic pole shifted position.
   b. there were two north magnetic poles over much of Earth's history.
   c. the north magnetic pole was fixed and the continents moved.
   d. none of these

7. In a convection cell _________.
   a. hot, less dense material rises upward
   b. hot, denser material rises upward
   c. cool, less dense material sinks downward
   d. cool, denser material rises upward

8. The magnetic anomalies about a ridge crest are due to
   a. alternating basalt and rhyolite flows.
   b. thick accumulations of magnetite rich sediment.
   c. manganese nodules.
   d. reversing polarity of the Earth's magnetic field.

9. Compared to oceanic crust, continental crust is
   a. younger.
   b. older.
   c. about the same age.
   d. cannot accurately be dated.

10. The newest oceanic crust and the thinnest deep-sea sediments are found at
   a. deep sea trenches.
   b. ridge crests.
   c. transform faults.
   d. abyssal plains.

11. The "supercontinent cycle" predicts the formation of a supercontinent roughly every ________ year.
   a. 5 million
   b. 50 million
   c. 500 million
   d. 5 billion

12. Convergent plate boundaries form where
   a. two continent bearing plates collide.
   b. two ocean bearing plates collide.
   c. an oceanic and a continental bearing plate collide.
   d. any of the above

13. Transform plate boundaries
   a. occur when two plates slide horizontally past one another.
   b. are exemplified by the San Andreas Fault.
   c. can offset ridge crests.
   d. all of the above

14. Trenches are NOT present at _____ convergent margins.
    a. ocean-ocean
    b. ocean-continent
    c. continent-continent
    d. any

15. The chaotic mixture of rock known as *melange* is indicative of a _______ plate boundary.
    a. divergent
    b. convergent
    c. transform
    d. any of these

16. The fastest moving plate is
    a. the Pacific plate.
    b. the North American Plate.
    c. the South African Plate.
    d. the Arabian Plate.

17. The Aleutian Islands, the Tonga arc, the Japanese and the Philippine Islands are an example of
    a. an oceanic and a continental bearing plate colliding.
    b. two continental bearing plates colliding.
    c. two ocean bearing plates colliding.
    d. hot spot formed island chains.

18. The Peru-Chile Trench and the Andes Mountains are examples of
    a. an oceanic and continental bearing plate colliding.
    b. two continental bearing plates colliding.
    c. two ocean bearing plates colliding.
    d. a transform fault.

19. The Himalayas, Appalachians and Alps are examples of mountain ranges formed
    a. where an oceanic and continental bearing plate collided.
    b. where two continental bearing plates collided.
    c. where two ocean bearing plates collided.
    d. along a transform fault.

20. Most of the worlds trenches border
    a. the Atlantic Ocean.
    b. South America.
    c. the Indian Ocean.
    d. none of the above

21. The Hawaiian Island chain is an example
    a. an oceanic and a continental bearing plate colliding.
    b. two continental bearing plates colliding.
    c. two ocean bearing plates colliding.
    d. a hot spot formed island chain.

22. As one moves away from a hot spot, islands tend to get _____ and ______.
    a. younger / larger
    b. younger / smaller
    c. older / larger
    d. older / smaller

23. The theory of plate tectonics was developed by
    a. Darwin.
    b. Einstein.
    c. many scientists including Wilson, Wegener, and Hess.
    d. Buckland.

24. To drill all the way through a tectonic plate, you would
    a. need to go down over 1000 km.
    b. need to go down over 100 km.
    c. have to drill through giant floating diamonds and amethyst caves.
    d. need to drill through multiple molten layers.

25. The fewest impact craters are seen on
    a. Mars.
    b. Mercury.
    c. Earth.
    d. Io.

TRUE OR FALSE

___1. Knowledge of plate tectonics can predict where earthquakes, volcanoes, and mineral resources will occur.

___2. Only the Northern Hemisphere continents formed the continent of Pangaea.

___3. When reconstructing Pangaea, the best reconstruction is given by matching present day shorelines.

___4. Glossopteris flora is found in all of the Southern Hemisphere continents.

___5. Sea floor spreading is a process in which sea floor forms at a ridge crest and is destroyed in a trench.

___6. In plate tectonics, the continents move laterally by pushing their way through the oceanic crust.

___7. Batholiths commonly occur at ocean-continent convergent margins.

___8. The highest mountains in the world are at ocean-continent convergent margins.

___9. The symmetrical pattern of magnetic anomalies about the axes of trenches, finally proved sea floor spreading.

___10. With increasing distance from a ridge crest, the sea floor becomes older and the sediments become thicker.

___11. Hawaii is an example of an island arc.

___12. The East African Rift Valley illustrates a continent in the early stages of fragmentation.

___13. When an ocean bearing plate collides, with a continent bearing plate, the ocean bearing plate always subducts beneath the continent bearing plate.

___14. Although back-arc basins are associated with convergent margins, they are under extensional stress and may be sites of oceanic crust formation.

___15. Transform boundaries tend to have little or no earthquake activity.

____16. Hot spots originate at plate margins.

____17. Plate velocity can be determined using magnetic anomalies

____18. Most geologists now agree that the main mechanism of plate movement is the "ridge-push".

____19. As of yet, there has been no demonstrable relationship between mineral resources and plate tectonics.

____20. The Atlantic Ocean is currently shrinking.

____21. If the continents had never moved, paleomagnetic data would indicate that the Earth has had multiple, moving magnetic poles.

____22. Magnetic reversals are exceedingly rare in the Earth's past.

____23. Ancient transform boundaries are easy to recognize because of the distinctive sequence of rocks they create.

____24. Gravity helps drive the plates through slab pull and ridge push.

____25. Hot spots can be used to calculate plate motion velocity and direction.

DRAWINGS AND FIGURES

1. For each plate boundary shown on the map below, indicate by arrows which are convergent and which are divergent. Draw arrows pointing toward each other at convergent boundaries, and away from each other at divergent boundaries.

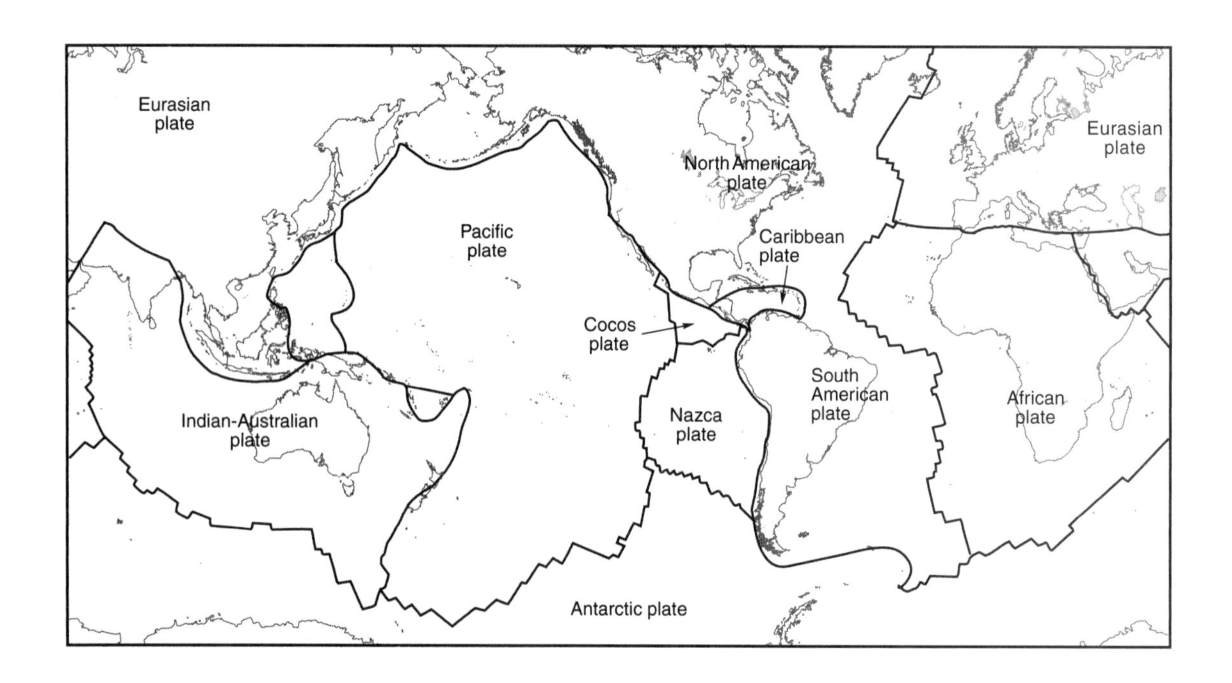

2. What type(s) of plate margin(s) is(are) shown in the block diagrams below? Label (a) a rift valley, (b) the continental seaboard, (c) fault blocks, and (d) the mid-ocean ridge.

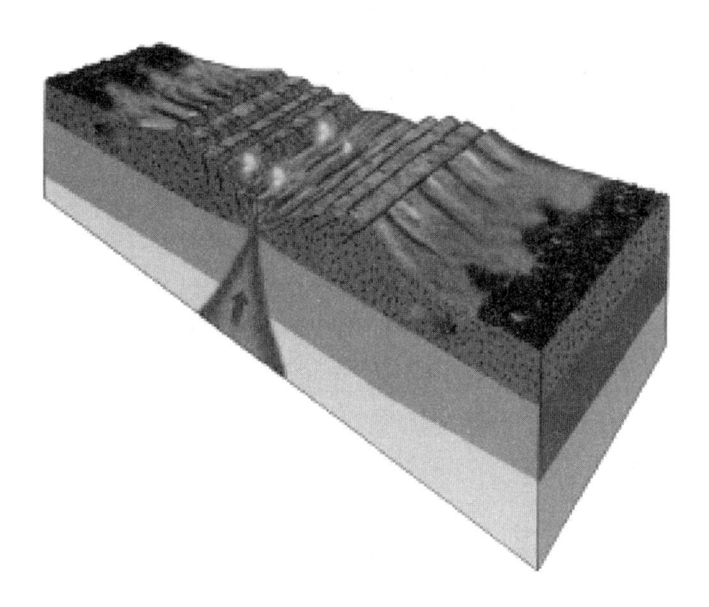

3. What types of plate margins are shown in the block diagrams below? Label the (a) volcanic arc(s), (b) the trench(es), (c) the back-arc basin(s), and (d)the subduction zone(s)

(a)

(b)

Chapter 3

Minerals—The Building Blocks of Rocks

CHAPTER OBJECTIVES

By the end of this chapter you should be able to:

1. Describe the parts of an atom and explain why some tend to ionize.
2. Describe the various types of bonds found in minerals.
3. Fully define "mineral" as defined by geologists and, for a variety of substances, say whether or not they are minerals and why.
4. List the common elements of Earth's crust in order of abundance.
5. List the common mineral groups and their characteristic compositions.
6. Explain how silicate minerals are put together.
7. Be familiar with the names of some of the most common minerals and be able to place them in the proper groups.
8. List and define several physical properties of minerals and know what controls them.
9. Explain a number of ways in which minerals form.
10. Differentiate between a rock and a mineral.
11. Describe several ways in which minerals have artistic and economic value.

A USEFUL ANALOGY

A simple model illustrating the effect of bond type (and, therefore, strength) on cleavage can be built by pasting several pencils together using classroom type paste. The pencils represent a chain of silica tetrahedra covalently bonded. The paste can be considered ionic bonds holding the chains together. Make several layers in this way. If you drop the model it will always break the pasted bonds (i.e. it will always break the bonds holding the pencils together) rather than breaking the pencils themselves. Although it is possible to break the pencils (the covalent bonds) it is much easier to break the pencils apart. Furthermore, when you break the pencils apart they break along planes (cleavage!) and when you break across the pencils the break is irregular (a lack of cleavage!).

KEY TERMS

After reading and studying this chapter you should know the following terms:

mineral
gemstones
elements
atoms
nucleus
protons
neutrons
electrons
electron shells
atomic number

atomic mass number
isotopes
bonding
compound
noble gas
ion
ionic bond
covalent bond
metallic bond
van der Waals bond
naturally occurring
inorganic
crystalline
crystal
amorphous
interfacial angle
native elements
radicals
silica
silicates
silica tetrahedron
ferromagnesian
nonferromagnesian
carbonate minerals
luster
crystal form
cleavage
fracture
hardness
Mohs hardness scale
specific gravity
density
double refraction
rock
rock-forming minerals
accessory minerals
resource
reserve
non-renewable

CHAPTER CONCEPT QUESTIONS

1. Completely define the term *mineral.*

2. How are atomic number and mass of an element determined?

3. Distinguish between an element and an isotope in terms of the nucleus of an atom.

4. How and why do many atoms tend to become ions?

5. Describe several ways in which atoms join together in compounds?

6. Explain why each of the following is or is not a mineral: quartz, glass, bone, water, ice.

7. Why don't minerals usually show well developed crystal forms?

8. What is meant by "constancy of interfacial angles"?

9. Look at Figure 3.10. In minerals it is common for calcium ions to substitute for sodium ions and for magnesium ions to substitute for iron ions without substantially changing the crystal structure. However, iron never substitutes for calcium or sodium. Why?

10. Considering the large number of elements in existence and number of combinations possible, why is Earth composed of relatively few abundant minerals?

11. Explain the difference between *silicon, silica*, and *silicates.*

12. How does the formula of the silica radical [$(SiO_4)^{-4}$, $(SiO_3)^{-2}$, SiO_2, etc.] in a mineral's chemical formula reflect the arrangement of the silica tetrahedra in that mineral?

13. List some characteristics that are used to identify minerals. For each, describe factors (composition, crystal structure, etc.) that determine that particular characteristic.

14. Why is it uncommon for minerals to display perfect crystal forms. What factors allow a crystal of a mineral to display its perfect crystal form?

15. What does a specific gravity of 2.3 mean?

16. Describe three ways in which minerals form.

17. Under what circumstances is a mineral a rock-forming mineral?

18. What is the difference between mineral resources and mineral reserves?

19. What factors other than geologic processes can cause an occurrence of a resource to be considered part of a nations reserves one year but not the next or vice versa?

COMPLETION QUESTIONS

1. To be considered a mineral, a substance must be _________ occurring and be a _________ solid (i.e. it must have a specific internal structure).

2. _________ are aggregates of one or more minerals.

3. Anything that has mass and occupies space is considered _________.

4. All matter is composed of chemical _________, which are composed of particles called _________.

5. The central portion of an atom is called the _________, in which there are positively charged _________ and neutrally charged _________.

6. The number of protons in the nucleus of an atom determines its _________ number.

7. The atomic mass number is the sum of the number of _________ and _________ present.

8. Isotopes of an element all have the same number of _________, but a different number of _________ in the nucleus.

9. The process of joining atoms together is called _________.

10. When atoms of two or more different elements are joined together the resulting substance is a _________.

11. When an atom gains or loses an electron, a charged particle called an _________ forms.

12. The attractive force between ions of opposite electrical charges produces _________ bonding. When adjacent atoms share electrons, _________ bonding occurs.

13. When the electrons of the outermost shell move freely from one atom to another, the atoms have _________ bonding.

14. Minerals that are only composed of one element are called _________.

15. Minerals are classified into groups that have the same negatively charged _________.

16. The two most abundant elements in Earth's crust are _________ and _________, which combine with other elements to form the group of minerals known as the _________.

17. The ferromagnesian silicates form _________ colored minerals.

18. The non-ferromagnesian minerals notably lack the elements ________ and ________ in their chemical make-up.

19. Calcite is one of the most abundant non-silicate minerals. It belongs to the _________ group.

20. Minerals are most commonly identified by using their _________ properties.

21. The appearance of a mineral in reflected light determines its _________.

22. The ability of a mineral to break along smooth planes is called _________ but when a mineral breaks along random, irregular surfaces it is showing _________.

23. The resistance of a mineral to abrasion determines its _________.

24. Specific gravity is the ratio of the weight of a mineral compared to the weight of an equal volume of _________.

25. The most abundant minerals that make up the bulk of Earth's crust are called the _________ forming minerals. Most belong to the _________ group of minerals.

MULTIPLE CHOICE

1. Minerals are
   a. naturally occurring.
   b. inorganic in their composition.
   c. crystalline solids.
   d. all of the above.

2. The smallest unit of matter that retains the characteristics of an element is an
   a. ion.
   b. isotope.
   c. atom.
   d. electron.

3. Protons have an electric charge that is
   a. negative.
   b. positive.
   c. neutral.
   d. none of the above

4. The atomic number is determined by the
   a. number of neutrons present.
   b. number of electrons present.
   c. number of protons present.
   d. number of protons plus the number of electrons present.

5. Protons are
   a. responsible for attracting and holding electrons.
   b. found only in metallic elements.
   c. negatively charged.
   d. found only in plasmas.

6. An element that has a full outer electron shell when electrically neutral is called a(n)
   a. ion.
   b. isotope.
   c. noble gas.
   d. radical.

7. A weak attractive force between electrically neutral atoms results in
   a. covalent bonds.
   b. ionic bonds.
   c. van der Waals bonds.
   d. metallic bonds.

8. Strong bonds in which electrons are shared between adjacent atoms are
   a. ionic.
   b. covalent.
   c. metallic.
   d. van der Waals.

9. If a mineral has a high electrical conductivity it most likely has a(n) ___ bond.
   a. ionic
   b. covalent
   c. metallic
   d. van der Waals

10. If a substance is constructed of atoms arranged in a regular three-dimensional arrangement it is
   a. crystalline.
   b. a glass.
   c. amorphous.
   d. none of the above

11. Minerals composed of atoms of a single type are called
   a. oxide minerals.
   b. silicate minerals.
   c. carbonate minerals.
   d. none of these.

12. The most abundant element in Earth's crust both by volume and mass is
   a. oxygen.
   b. hydrogen.
   c. iron.
   d. silicon.

13. In the classification scheme used for minerals, all minerals are divided into seven groups based on
   a. physical characteristics.
   b. the positively charged ions in the formula.
   c. the negatively charged ions in the formula.
   d. crystal shape.

14. Tightly bonded groups of atoms (e.g. CO_3^{-2}, SO_4^{-2}, and OH^{-1}) that behave as a single unit in a mineral are called
   a. isotopes.
   b. radicals.
   c. noble gasses.
   d. compounds.

15. The group of minerals that contains the largest number of members by far is the _________ group.
    a. carbonate
    b. sulfate
    c. silicate
    d. sulfide

16. The charge on a silica tetrahedron unlinked to any other silica tetrahedra is
    a. +2.
    b. –2.
    c. –4.
    d. +4.

17. Ferromagnesian silicates are rich in
    a. iron and manganese.
    b. iron and magnesium.
    c. silicon and aluminum.
    d. none of the above

18. The most common minerals in Earth's crust belong to a group known collectively as
    a. quartz.
    b. biotite.
    c. augite.
    d. feldspar.

19. Quartz is composed of
    a. calcium carbonate.
    b. silicon and oxygen.
    c. potassium and silicon.
    d. none of these

20. A very abundant carbonate mineral is
    a. gypsum.
    b. halite.
    c. calcite.
    d. hematite.

21. When a mineral breaks along smooth planes, it has
    a. fracture.
    b. streak.
    c. cleavage.
    d. all of these

22. Which physical property is controlled by structure and bond strength?
    a. color
    b. luster
    c. cleavage
    d. specific gravity

23. There a limited variety of minerals on Earth because
    a. most elements are rare.
    b. only eight elements are common in Earth's crust.
    c. many combinations of elements don't occur.
    d. all of these answers

24. Chemistry, the interaction of atoms, is a function of
    a. the number of neutrons.
    b. the size of the protons.
    c. the electrons.
    d. the nuclear stability of the atom.

25. Mineral resources are
    a. most common in industrial countries.
    b. most common in non-industrial countries.
    c. easily replaceable.
    d. nonrenewable.

TRUE OR FALSE

____1. Minerals are assemblages of rocks.

____2. Although the nucleus makes up only a tiny fraction of an atom's volume, it contains almost all of the mass.

____3. Electrons have a negative charge.

____4. The atomic mass number is determined by the sum of the number of protons and the number of neutrons present.

____5. The atomic number of an element may vary from atom to atom but the atomic mass number may not.

____6. Covalent bonding is due to the sharing of electrons.

____7. Minerals are always crystalline solids.

____8. If a mineral doesn't show well-developed crystal faces it is not crystalline.

____9. Two crystals of the same mineral will generally exhibit identical angle between similar crystal faces, even if they are of very different sizes.

____10. All minerals have specific chemical formulas that cannot vary.

____11. Silica is composed of silicon and carbon.

____12. The basic building block of the silicate minerals is tetragonal in shape.

____13. Nonferromagnesian silicates are dark in color.

____14. Quartz will scratch glass.

____15. Calcite is the most common carbonate mineral.

____16. Hematite and magnetite are very important ores of copper.

____17. Color is the most diagnostic of all physical properties.

____18. Fracture is mineral breakage along random, irregular surfaces.

___19. All minerals have cleavage.

___20. Ferromagnesian minerals generally have a higher specific gravity than non-ferromagnesian minerals

___21. Abundance is an important criterion for characterizing a mineral as *rock forming*.

___22. A resource is the total amount of a commodity whether discovered or undiscovered.

___23. A resource does not necessarily have to be extractable to be considered a reserve.

___24. Most mineral resources are renewable.

___25. The U.S. must import a large number of mineral resources primarily because it contains so few naturally.

DRAWING QUESTION

1. Sketch the structure of olivine, pyroxene, amphibole, and mica.

2. Draw and explain the how to tell the difference between the cleavage of halite and that of calcite.

Chapter 4

Igneous Rocks and Intrusive Igneous Activity

CHAPTER OBJECTIVES

By the end of this chapter you should be able to:

1. Explain the difference between felsic, intermediate, and mafic composition of rocks or magma.
2. Know what viscosity is, what controls it, and be able to explain how it affects the behavior of magma.
3. Reproduce Bowen's Reaction series and explain how it works.
4. Explain how magma is generated at mid-ocean ridges
5. Suggest reasons why magmas associated with subduction zone volcanism are commonly intermediate to felsic in composition.
6. Describe three ways in which an initially mafic magma can become more silica rich.
7. List and define the common textures found in igneous rocks.
8. Classify the igneous rocks using composition and texture.
9. Classify the various types of igneous intrusions by their geometry and relationships to the surrounding country rock.
10. Explain why the large volume of granite in Earth's crust presents a problem for a strictly igneous origin.
11. Be familiar with the various theories for emplacement of batholiths.

A USEFUL ANALOGY

It may not be immediately obvious why the crystallization and subsequent removal of olivine from magma via crystal settling enriches the remaining magma in silica. Here is a simple mathematical experiment to show that it happens. Although far simpler than an actual magma, the principle is the same:

Imagine a jar filled with 100 marbles of three different colors. Fifty of the marbles are blue, 25 are green, and 25 are red. This is analogous to magma with 50% silica (blue marbles), 25% magnesium (green), and 25% iron (red). To make an olivine you combine one of each color. To remove that olivine from the magma, simply remove one marble of each color from the jar. Do this ten times then calculate how many of each color you have left. You should have 40 blue, 15 green, and 15 red for a total of 70 marbles. Whereas initially 50% of the marbles were blue, after the removal of "olivine" (composed of green and red marbles), blue marbles constitute 57% of the total. The jar is thus enriched in blue marbles (i.e. the silica)!

KEY TERMS

After reading and studying this chapter, you should know the following terms:

igneous rock
magma
lava
pyroclastic
extrusive
intrusive

plutonic
felsic
intermediate
mafic
viscosity
magma chamber
Bowen's reaction series
continuous branch
discontinuous branch
geothermal gradient
mantle plumes
partial melting
crystal settling
assimilation
country rock
inclusion
magma mixing
texture
aphanitic
phaneritic
porphyritic
phenocrysts
groundmass
natural glass
vesicle
vesicular
pyroclastic or fragmental texture
ultramafic
peridotite
basalt
gabbro
andesite
diorite
rhyolite
granite
granodiorite
granitic
pegmatite
ash
tuff
welded tuff
volcanic breccia
obsidian
pumice
pluton
concordant
discordant
dike
sill
laccoliths
volcanic pipe
volcanic neck
batholith
stock
granitization
forceful injection
stoping

CHAPTER CONCEPT QUESTIONS

1. How does an extrusive igneous rock differ from an intrusive igneous rock in mode of origin and in texture?

2. What is the difference between the continuous and the discontinuous branches of Bowen's reaction series?

3. Why does crust melt below spreading ridges to form magma?

4. Why does the magma that forms as oceanic crust at spreading ridges have a mafic composition even though the mantle that melts to form it is ultramafic?

5. Explain two ways in which a mafic magma can become richer in silica as it moves towards the surface.

6. On what basis are igneous rocks classified? How do geologists generally identify igneous rocks?

7. Why are rhyolite lava flows so rare compared to andesite and basalt lava flows?

8. Why does pegmatite tend to grow relatively few large crystals rather than many smaller crystals like most igneous rocks? What accounts for the large number of unusual minerals in some pegmatites?

9. Pumice is both vesicular and glassy. What does this tell you about the magma from which it formed?

10. Explain the difference between a dike, sill, laccolith, batholith and stock.

11. Why do geologists believe that some granite may be of metamorphic origin? How do they distinguish those granites from those that are intrusive?

12. What has the study of salt domes in sediments beneath the Gulf Coast suggested about the emplacement mechanism of plutons?

13. How does batholith emplacement by stoping differ from emplacement by forceful injection?

COMPLETE THE FOLLOWING TABLE:

	Felsic	**Intermediate**	**Mafic**	**Ultramafic**
% Silica				
Aphanitic rock				
Phaneritic rock				

COMPLETION QUESTIONS

1. Based on their origin, there are two types of igneous rocks, the volcanic or _________ igneous rocks, and the plutonic or _________ igneous rocks.

2. In the discontinuous branch of Bowen's reaction series, as the magma cools, _________ rich plagioclase feldspar reacts with the melt to form _________ rich plagioclase feldspar.

3. The discontinuous branch of Bowen's reaction series begins with the crystallization of _________ at high temperatures and ends with the crystallization of _________ .

4. Temperature increases at a rate of about _____ °C/km going downward into the crust.

5. Crystal settling, assimilation and magma mixing are processes that, to some degree, change the composition of magma from relatively _________ in silica to relatively _________ in silica.

6. Volcanic igneous rocks have crystals that are too fine to be seen with the naked eye and are said to have a_________ texture.

7. Plutonic igneous rocks have crystals coarse enough to be seen with the naked eye and are said to have a _________ texture.

8. If an igneous rock has experienced two phases of cooling, one slowly at depth and a second more quickly on the surface, then it develops a _________ texture.

9. An extrusive rock with holes, due to escaping gases, has a _________ texture.

10. Explosive volcanic eruptions produce fragments of various sizes that accumulate to form rocks with a _________ texture.

11. Igneous rocks are classified according to their _________ and mineralogical _________.

12. Felsic rocks have a silica content exceeding ____%.

13. A felsic, aphanitic igneous rock is called a _________.

14. An intermediate, phaneritic igneous rock is called a _________.

15. _______ is the aphanitic compositional equivalent of gabbro.

16. The most abundant mineral in ultramafic rocks is _________.

17. An igneous rock in which most crystals exceed one cm. in size is called a _________.

18. Most pegmatites are similar to _________ in composition.

19. A rock composed of fine-grained ash particles is classified as a _________.

20. ________ is generally a dark colored volcanic glass with a conchoidal fracture, whereas _________ is a vesicular volcanic glass.

21. When magma solidifies in the subsurface, it forms a rock body called a _________.

22. A tabular, discordant pluton is called a ________ and a tabular, concordant one is a _________.

23. Non-tabular, concordant plutons are called _________.

24. _________ is a mechanism of batholith emplacement where blocks of country rock are engulfed by, rather than shouldered aside by rising magma.

25. Most batholiths are _________ in composition, although a few are _________ in composition.

MULTIPLE CHOICE

1. Intrusive igneous rock forms
   a. on the surface.
   b. by violent volcanic eruptions.
   c. from glassy lavas.
   d. none of these.

2. Compared to felsic magmas, mafic magmas are relatively enriched in
   a. calcium.
   b. iron.
   c. magnesium.
   d. all of the above

3. The discontinuous branch of Bowen's reaction series comprises the
   a. ferromagnesian silicates.
   b. nonferromagnesian silicates.
   c. highest silica minerals.
   d. all of these

4. The discontinuous branch of Bowen's reaction series comprises
   a. silicates made of isolated silica tetrahedra.
   b. single chain silicates.
   c. double chain silicates.
   d. all of the above

5. The continuous branch of Bowen's reaction series consists of plagioclase feldspars in which calcium ions are replaced by ____ ions during cooling
   a. potassium.
   b. magnesium.
   c. iron.
   d. none of the above

6. The last mineral to form as a magma solidifies is
   a. olivine.
   b. quartz.
   c. feldspar.
   d. muscovite.

7. Magmas formed at spreading ridges are invariably
   a. felsic.
   b. intermediate.
   c. mafic.
   d. ultramafic.

8. Partial melting of mantle or mafic crust explains
   a. the age of some igneous rocks.
   b. the size of some volcanoes.
   c. the chemical composition of some magmas.
   d. none of the above

9. The presence of inclusions of country rock in a felsic intrusive igneous rock may be evidence of
   a. crystal settling.
   b. assimilation.
   c. magma mixing.
   d. all of the above

10. Coarse-grained or phaneritic rocks indicate
    a. slow cooling on the surface.
    b. rapid cooling at depth.
    c. slow cooling at depth.
    d. none of these

11. The presence of phenocrysts in an igneous rock indicate
    a. cooling so rapid that crystals do not have time to form.
    b. violent, explosive volcanic eruption.
    c. the escape of gas from a magma.
    d. two phases of cooling, one fast and one slow.

12. Vesicles in an igneous rock form from
    a. escaping gases.
    b. phenocrysts.
    c. falling ash.
    d. none of these

13. A glassy texture indicates
    a. very rapid cooling.
    b. slow cooling.
    c. slow cooling and rapid cooling.
    d. none of these

14. Pyroclastic igneous rocks from by
    a. a lava cooling on the surface.
    b. violent, explosive volcanic eruption.
    c. a magma cooling slowly in the subsurface.
    d. two phases of cooling, one fast and one slow.

15. An aphanitic, dark-colored igneous rock is called a
    a. basalt.
    b. gabbro.
    c. rhyolite.
    d. none of these

16. Rhyolite is a(n)
    a. phaneritic intermediate igneous rock.
    b. aphanitic, mafic igneous rock.
    c. phaneritic, felsic igneous rock.
    d. aphanitic, felsic igneous rock.

17. By far most intrusive igneous rocks are _________ in composition.
    a. granitic
    b. andesitic
    c. basaltic
    d. All of the above are found in equal abundance.

18. Which of the following is a characteristic of pegmatites?
    a. very large grain size
    b. gem minerals
    c. occurrence in granitic rocks
    d. all of these

19. Which of the following rocks has a glassy texture?
    a. granite
    b. obsidian
    c. tuff
    d. basalt

20. Which of the following rocks is pyroclastic?
    a. basalt
    b. tuff
    c. rhyolite
    d. andesite

21. Plutons form when
    a. magma solidifies in the subsurface.
    b. lava cools on the surface.
    c. a volcano erupts.
    d. none of these

22. Dikes are
    a. tabular in shape and oriented parallel to the surrounding rock layers.
    b. mushroom shaped.
    c. tabular in shape and crosscut the surrounding rock layers.
    d. none of these

23. Batholiths are found in association with
    a. plateau basalts.
    b. oceanic islands.
    c. mountain ranges.
    d. the oceanic crust.

24. Granitization is the formation of granite by extreme _________ of country rock.
    a. metamorphism
    b. weathering
    c. injection
    d. assimilation

25. Felsic lava is less common than mafic because
    a. mafic lava is too tough to erode.
    b. felsic magma is more viscous.
    c. mafic magma is more viscous.
    d. felsic lava erodes faster.

TRUE OR FALSE

____1. Extrusive igneous rocks form from magma cooling in the subsurface.

____2. Felsic igneous rocks are richer in silica than mafic igneous rocks.

____3. Magma generally cools quickly because rock is an excellent conductor of heat.

____4. The higher the viscosity of lava, the faster it will flow.

____5. A felsic magma has a lower viscosity than a mafic magma.

____6. Most magma originates from the Earth's molten core.

____7. As an ultramafic magma cools, the first minerals to form are olivine and calcium plagioclase.

____8. The discontinuous series consists of nonferromagnesian minerals.

____9. The geothermal gradient is due to a decrease in temperature with depth.

____10. Different minerals have different melting points.

____11. If a mafic magma of a given volume undergoes crystal settling, it will produce an equal volume of felsic magma.

____12. Assimilation refers to two magmas mixing together.

____13. The texture of an igneous rock is determined by the cooling history of the magma.

____14. The mantle is believed to be composed primarily of basalt.

____15. Ultramafic lavas have been particularly abundant over the last 100 million years.

____16. The richer a magma is in silica, the lighter in color it will generally be.

___17. It would not be unusual to see olivine phenocrysts in a porphyritic basalt.

___18. Basalt is the most common extrusive igneous rock.

___19. Abundant water vapor in the late stages of magma crystallization inhibits nucleation of crystals.

___20. Obsidian is a common constituent of intrusive igneous rocks.

___21. Granite is exposed on the surface only after uplift and erosion of the overlying rocks.

___22. Laccoliths and sills have the same shape.

___23. Batholiths are the largest of all plutons.

___24. Granites can be both igneous and metamorphic in origin.

___25. The presence of inclusions of country rock in a granite may indicate emplacement by stoping.

DRAWINGS AND FIGURES

1. What two types of plutons are shown in the figure below?

Courtesy Frank Hanna

2. Using the diagram below, identify aphanitic and phaneritic rocks with the following compositions:
   a. 5% quartz, 38% potassium feldspar, 24% plagioclase feldspar, 15% biotite, 8% hornblende.
   b. 20% plagioclase, 8% hornblende, 56% pyroxene, 16% olivine
   c. 60% plagioclase, 1% biotite, 31% Hornblende, 8% pyroxene

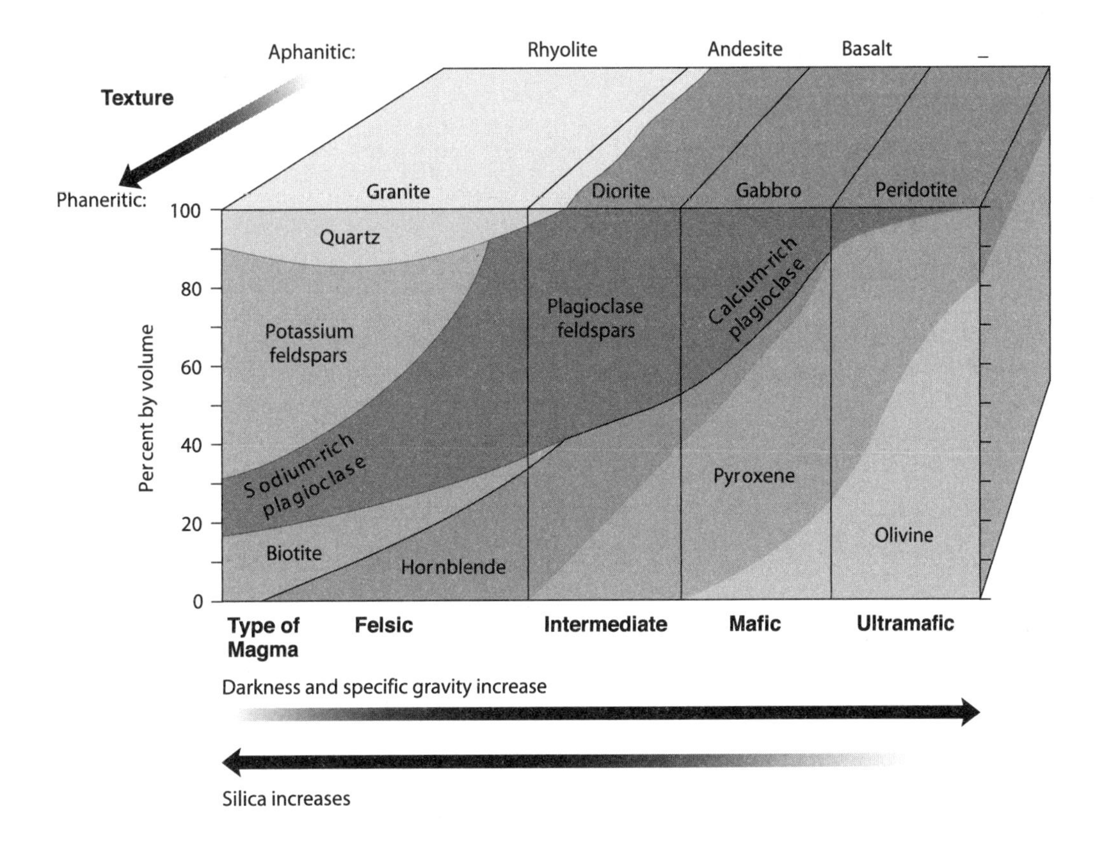

3. Sketch cross-sectional diagrams of the following plutons intruded into sedimentary (i.e. layered) country rock:
   a. stock
   b. sill
   c. laccolith
   d. dike

Chapter 5

Volcanoes and Volcanism

CHAPTER OBJECTIVES

By the end of this chapter you should be able to:

1. Differentiate between active, dormant, and extinct volcanoes.
2. Be familiar with the different gasses, lavas, and pyroclastic materials released during volcanic eruptions.
3. Explain how a caldera forms.
4. List the major types of volcanoes and describe the differences among them.
5. Characterize the deposits of fissure eruptions and pyroclastic eruptions.
6. Discuss the current state of volcano eruption monitoring and prediction both in terms of techniques employed and success of predictions.
7. Describe the locations of several volcanic belts around the world and explain in a plate tectonic context why they occur where they do.
8. Be able to characterize the types of volcanic activity that occur in various plate tectonic settings.

A USEFUL ANALOGY

Why does viscosity affect the type of eruption that a volcano will have? Think back to your childhood when you wouldn't have considered drinking milk with a straw without blowing bubbles in it. It was pretty easy to do; you didn't have to blow too hard. But how about when you drank a milkshake? That was tougher. You might have had to blow until red in the face. When you finally blew hard enough the milkshake likely exploded out of the cup. The reason is that the "thicker" milkshake required more pressure to get it out of the way of the rising gas (your breath). A magma that has a high viscosity similarly requires a high gas pressure to get it to move out of the way. Gas builds until that pressure is finally reached. Then the rapid release of pressure may result in an explosive eruption.

KEY TERMS

After reading this chapter, you should be able to define the following terms:

volcanism
active volcano
dormant volcano
extinct volcano
vog
lava tube
pahoehoe
aa
pressure ridge
spatter cones
columnar joints
pillow lava
ash
ash fall
ash flow
lapilli

volcanic bomb
volcanic block
volcano
crater
caldera
shield volcano
Hawaiian-type eruptions
cinder cone
composite volcano (statovolcano)
lahars
lava dome
nuee ardente
fissure eruption
basalt plateau
pyroclastic sheet deposit
welded tuff
Volcanic Explosivity Index
volcanic or harmonic tremor
circum-Pacific belt
Mediterranean belt

CHAPTER CONCEPT QUESTIONS

1. What are the differences between active, dormant and extinct volcanoes?

2. What are the common gasses associated with volcanism and what problems can they cause?

3. How does viscosity affect the characteristics of a lava flow?

4. Explain the origin of each of the following volcanic features:
   a. spatter cone
   b. columnar jointing
   c. pressure ridge
   d. pillow lava
   e. caldera
   f. lava dome
   g. welded tuff
   h. basalt plateau

5. How does lava flowing down a valley ultimately produce a topographic ridge?

6. What type of eruptions and what composition of magma produce shield volcanoes? Why?

7. What type of eruptions and what composition of magma produce cinder cones?

8. What types of deposits are found associated with composite volcanoes?

9. Describe four famous volcanoes or volcanic features in the Cascade Range.

10. What is the Volcanic Explosivity Index and how is it determined?

11. Describe several observations that are commonly made when monitoring a volcano for the purposes of predicting an eruption. How would you rate the success of these efforts in forecasting eruptions?

12. How does plate tectonics control the distribution of volcanoes? Relate different plate tectonic settings to the type of volcanism found there.

13. What do the ages of the Hawaiian Islands reveal about Pacific Plate movement?

COMPLETION QUESTIONS

1. A volcano which has not erupted in historic times but is expected to erupt again is said to be _______, whereas one which will probably never erupt again is _______.

2. By far, the most abundant gas associated with volcanism is ________.

3. Active volcanoes are found on at least two objects in the solar system: ________ and _______.

4. A lava flow may contract as it cools forming polygonal cracks called _______.

5. Basalts erupting under water form a distinctive type of lava called _______.

6. Finest-grained pyroclastic material (that less than 2 mm in diameter) is called _______.

7. Lava can be erupted from conical mountains called _______, or can be extruded from cracks in the Earth's crust called ________.

8. When a volcano erupts from its summit the lava comes out of a circular opening called a _______.

9. A _______ forms when the top of a volcanic mountain collapses into an empty magma chamber.

10. Big, broad volcanoes with gently sloping sides are called ________ volcanoes. Lavas from these volcanoes are usually _______ in composition.

11. The most continuous eruption in history is the ongoing eruption of _______ .

12. Small volcanoes composed of pyroclastic debris are called ________ cones.

13. Large, steeply sloping conical volcanoes composed of alternating layers of lava and pyroclastic debris are called ________ cone volcanoes.

14. Large volcanic mudflows called _______, may make up a large portion of composite cones.

15. The Hawaiian Islands are dominated by ______ volcanoes.

16. The Cascade Range of the Pacific Northwest is dominated by _______ cones.

17. Extensive but relatively thin sheets of basaltic lava erupted from long fissures are called _______.

18. Sheet-like deposit similar to those described in number 17, but made of material ejected by explosive eruptions are ________.

19. The possible range of the Volcanic Explosivity Index (VEI) is 0 to ____, but no historic eruption has exceeded a value of ____.

20. A continuous earthquake-like ground motion called _______ indicates moving magma below the surface of a volcano.

21. New oceanic crust is produced by volcanoes located along _______.

22. The circum-Pacific belt is a long belt of volcanoes located above ______ at convergent margins.

23. Volcanism at convergent plate boundaries is characterized by _______ volcanoes.

24. A cooling lava flow may crack and form _________ ________.

25. Mafic magma tends to erupt quietly because its ____ viscosity allows gases to _______ and ________.

MULTIPLE CHOICE

1. If a volcano has erupted in historic times it is considered
   a. active.
   b. dormant.
   c. extinct.
   d. none of the above

2. The odor and haze associated with the condition known as *vog* is due which gas?
   a. water vapor
   b. sulfur dioxide
   c. chlorine
   d. carbon monoxide

3. Lava that cools to form a pile of rough, jagged blocks is called
   a. pahoehoe.
   b. aa.
   c. nuee ardentes.
   d. lava tube.

4. The more viscous a lava, the more
   a. slowly it flows.
   b. likely it is to erupt violently.
   c. silica it contains.
   d. all of the above

5. When magma rises to the surface, gases are released because
   a. the pressure is reduced.
   b. the pressure is increased.
   c. the silica content changes.
   d. none of these

6. If tension cracks develop in a cooling lava, a _________ lava forms.
   a. blocky
   b. glassy
   c. columnar jointed
   d. none of these

7. Ejected globs of molten lava which cool during flight are called
   a. aa.
   b. volcanic blocks.
   c. volcanic bombs.
   d. nuee ardentes.

8. Shield volcanoes typically have slopes
   a. from 2 to 10 degrees.
   b. of around 30 degrees at the summit to as low as 5 degrees at the base.
   c. as steep as pyroclastic material will rest (typically about 33 degrees).
   d. nearly 90 degrees.

9. Mt. Pinatubo and Mt. St. Helens are examples of which type of volcano?
   a. shield volcano
   b. cinder cone
   c. composite volcano
   d. fissure volcano

10. Materials of _______ composition dominate most composite volcano eruptions.
    a. felsic
    b. intermediate
    c. mafic
    d. ultramafic

11. Crater Lake, Oregon is an example of
    a. a fissure.
    b. a volcanic crater.
    c. a caldera.
    d. none of the above

12. Lahars commonly make up large percentage of
    a. shield volcanoes.
    b. cinder cones.
    c. composite cones.
    d. fissures.

13. Eruptions of high volumes of low viscosity lava from long fissures result in
    a. basalt plateaus.
    b. pyroclastic sheets.
    c. lava domes.
    d. all of the above

14. The largest volcanoes on earth in terms of volume are
    a. shield volcanoes.
    b. cinder cones.
    c. composite cones.
    d. All of the above are about the same size.

15. The smallest but steepest type of volcano is
    a. shield volcanoes.
    b. cinder cones.
    c. composite cones.
    d. fissures.

16. A nuee ardente will most likely result in a
    a. pillow lava.
    b. welded tuff.
    c. basalt plateau.
    d. pressure ridge.

17. Which of the following aspects of a volcano are NOT monitored when trying to predict an eruption?
    a. changes in slope of the volcano's sides
    b. gas emissions
    c. volcanic tremors
    d. All of the above ARE monitored.

18. Which of the following is NOT found in the Cascades?
    a. composite cones
    b. shield volcanoes
    c. cinder cones
    d. All ARE found in the Cascades.

19. Mt. Rainier is considered by many to be America's most dangerous volcano. The main threat is likely to be
    a. collapse of the caldera.
    b. lahars.
    c. pyroclastic flows.
    d. lava flows.

20. Which of the following is NOT considered when assigning a Volcanic Explosivity Index (VEI) value to an eruption?
    a. number of people killed
    b. volume of material explosively ejected
    c. height of the ash plume
    d. All of the above ARE considered.

21. Most of the Earth's major volcanoes occur
    a. along the circum-Mediterranean belt.
    b. in the middle of plates.
    c. along the circum-Pacific belt.
    d. none of these

22. Submarine volcanic eruptions along mid-ocean ridges generally result in
    a. pillow basalts.
    b. pyroclastic sheet deposits.
    c. plateau basalts.
    d. lahars.

23. The circum-Pacific belt is characterized by volcanism
    a. along divergent plate margins.
    b. along convergent plate margins.
    c. along transform plate margins.
    d. in the interior of plates.

24. The Hawaiian Islands formed
    a. along divergent plate margins.
    b. along convergent plate margins.
    c. along transform plate margins.
    d. in the interior of plates.

25. Volcanic eruptions can affect climate by releasing
    a. sulfur dioxide that cools the atmosphere.
    b. dust that blocks sunlight.
    c. heat from lava and pyroclasts.
    d. all of these answers

TRUE OR FALSE

___ 1. A dormant volcano can become an active volcano.

___ 2. Carbon dioxide is the most abundant volcanic gas.

___ 3. Volcanic gasses alone cannot cause death if lava is not present.

___ 4. In the years following a large eruption the Earth generally experiences an increase in average global temperature.

___ 5. Pahoehoe can become aa as it flows away from the point of eruption.

___ 6. Lava flowing through lava tubes can generally flow faster and farther than lava flowing across the surface.

___ 7. Pillow lavas form when lava is extruded under water.

___ 8. Ash seldom travels far from the point of eruption.

___ 9. A caldera is relatively small compared to a crater.

___ 10. Crater Lake could be more accurately called "Caldera Lake".

___ 11. Shield volcanoes are composed of about 99% lava flows.

___ 12. Cinder cones are large steep cones composed of alternating pyroclastic materials, and lava flows.

___ 13. Low viscosity lava flows tend to flow down predictable paths and so usually present minimum danger to human life.

___ 14. Cinder cones often form on the flanks of larger shield volcanoes or composite cones.

___ 15. Cinder cones are generally not active as long as other types of volcanoes.

___ 16. Composite volcanoes generally have steeper slopes near the base than at the summit.

___ 17. Lava domes are usually mafic, but sometimes intermediate in composition.

___ 18. Although common, nuee ardentes rarely constitute a danger to human life.

___ 19. With the possible exception of Mt. St. Helens, the Cascade Volcanoes are thought to be extinct and unlikely to erupt again.

___ 20. The Columbia Plateau basalts are due to lavas extruded from composite cone volcanoes.

___ 21. Three quarters of the world's historic eruptions have lasted six months or less.

___ 22. In evaluating the eruptive potential of a volcano, the past eruptive history of the volcano is usually very deceptive and so is generally ignored.

___ 23. When looking at the world as a whole, volcanic eruptions are quite random.

___ 24. Most pyroclastic eruptions are from volcanoes located along spreading ridges.

___ 25. All of the Hawaiian Islands formed at roughly the same time.

DRAWINGS AND FIGURES

1. Draw a series of sketches illustrating the development of a caldera.

2. Sketch a cross-sectional view of a shield volcano, a composite cone, and a cinder cone labeling lava flows, pyroclastic materials, the central vent, craters and calderas as appropriate.

3. Use the graph below to determine the VEI of and eruption which:
   a. ejects a million cubic meters of tephra with a cloud height of 1-5 km.
   b. ejects .1 cubic kilometers of tephra with a cloud height of 3-15 km.
   c. ejects 50 cubic kilometers of tephra with a cloud height of over 25 km.

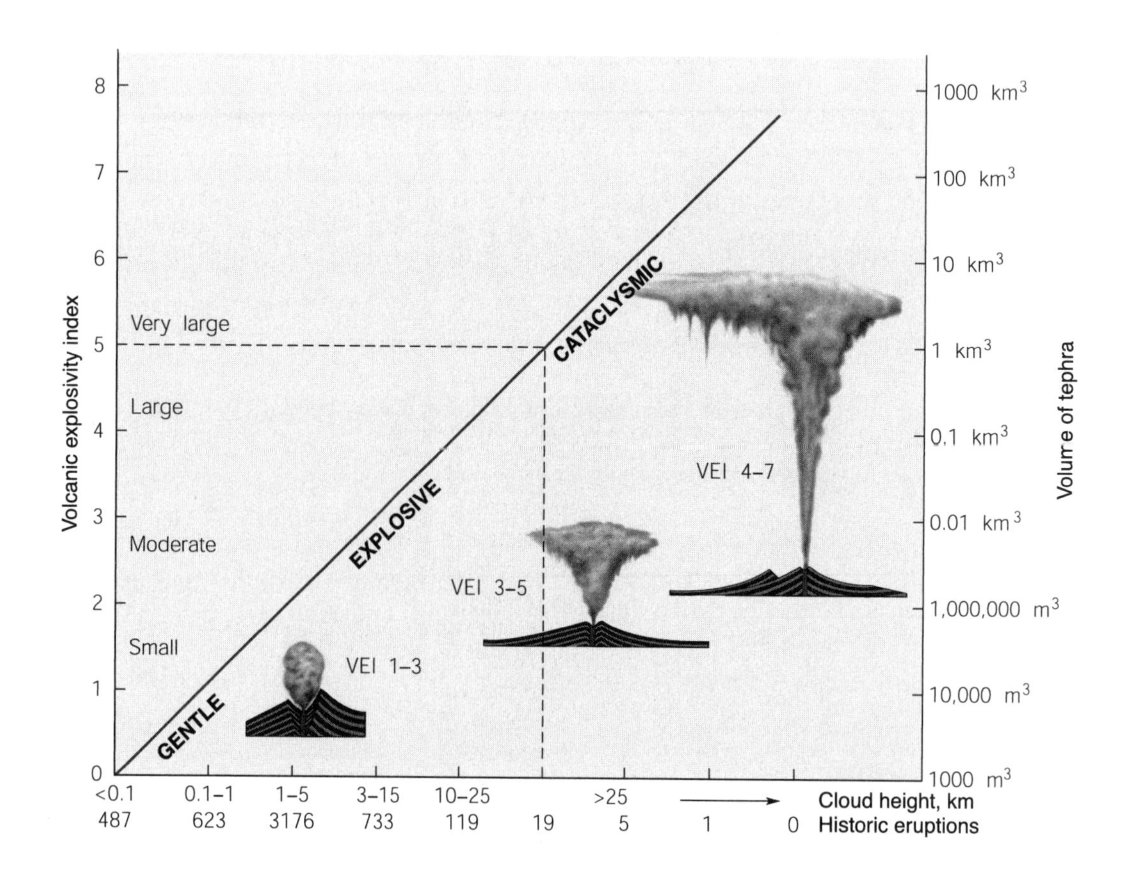

Chapter 6

Weathering, Erosion, and Soil

CHAPTER OBJECTIVES

By the end of this chapter you should be able to:

1. Differentiate between mechanical and chemical weathering.
2. List and explain six different mechanical weathering processes.
3. List and explain three different chemical weathering processes.
4. List and explain three different factors that control the rate of weathering.
5. List the essential components of soil and place them in the context of the soil profile.
6. List and explain five different factors that control rate and type of soil formed.
7. Describe several ways in which soil can be degraded.
8. List several mineral resources that are mined in the soil.

A USEFUL ANOLOGY

To see how surface area affects weathering rate dissolve a teaspoon of table salt in a glass or beaker. At the same time dissolve an equal volume of rock salt that you buy to thaw sidewalks in winter or put in water softeners. If the latter is unavailable in your part of the country you may have a similar sized piece of halite in your lab. The table salt dissolves before the coarser counterpart every time.

KEY TERMS

After reading this chapter, you should be familiar with the following terms:

weathering
parent material
erosion
transport
differential weathering
mechanical weathering
chemical weathering
frost action
frost wedging
talus
joints
frost heaving
pressure release
sheet joints
exfoliation
exfoliation domes
rock bursts
popping
thermal expansion and contraction
solution
carbonic acid
oxidation
hydrolysis
spheroidal weathering
regolith
soil

humus
residual soil
transported soil
loess
residual soil
transported soil
loess
soil horizons
topsoil
leaching
subsoil
zone of accumulation
pedalfer
pedocal
caliche
alkali soils
laterite
relief
expansive soils
soil degradation
sheet erosion
rill erosion
gullies
salinization
bauxite
residual concentration
gossan

CHAPTER CONCEPT QUESTIONS

1. Explain the difference between frost heaving and frost wedging.

2. How are exfoliation domes formed?

3. How does the excavation of deep, underground mines cause sheet joints to form? What danger do they pose miners?

4. How can forest fires act as a mechanical weathering agent?

5. Describe a way in which the activities of organisms aid in mechanical weathering and a way in which they participate in chemical weathering.

6. What happens at the molecular level when a mineral dissolves in water? When a mineral dissolves in a weak acid?

7. Why does weathering of limestone result in bold cliffs in arid climates and subdued, rolling topography in more humid areas?

8. How does clay form?

9. Describe a climatic condition that would promote chemical weathering. What characteristics of the parent material itself promote chemical weathering?

10. How does Bowen's reaction series predict susceptibility of minerals to chemical weathering?

11. Compare and contrast a soil found in the humid eastern U.S. with one found in the desert southwest and one found in equatorial Brazil.

12. Even though soil formed in tropical rainforest can be extremely thick, it is among the most unproductive for agriculture. Explain why.

13. Explain how parent rock type, climate, and slope affect amount and type of soil that develops.

14. Explain the differences between, and give examples of, erosion, physical soil degradation, and chemical soil degradation. For each example think of a way that farming practices can mitigate damage.

15. Explain why the richest aluminum ore deposits are products of weathering.

COMPLETION QUESTIONS

1. The physical breakdown and chemical alteration of rocks and minerals on the surface of the Earth is called _________.

2. The removal of weathered material is called _________.

3. The process of two different rocks weathering at two different rates is called _________ weathering.

4. The two types of weathering are _________ and ________ weathering.

5. The repeated freezing and thawing of water in cracks of rock produces ________ wedging.

6. Exfoliation domes are due to a release in ________ because overlying rocks have been ________ away.

7. Chemical weathering can not occur without the presence of ________.

8. Common chemical weathering reactions include ions going into _________, iron bearing minerals being ________, and feldspars undergoing ________.

9. Hydrolysis reactions on feldspar produces _________.

10. _________ size particles chemically weather fastest.

11. Initially rectangular blocks of rock subjected to chemical weathering tend to become more ________ as weathering proceeds.

12. The early-formed minerals of Bowen's reaction series weather ________ than the later formed minerals because they are lower in silica.

13. Naturally occurring surface materials that support plant life are called ________.

14. A soil derives its dark color from ________ matter.

15. Soils that form over transported and deposited sediment rather than bedrock are called ________ soils.

16. In a soil profile, the ________ horizon contains mineral and organic matter, while the ________ horizon is a clay-rich layer.

17. The A horizon of a soil is called the zone of ________, and the B-horizon is called the zone of ________.

18. The most important factor in determining rate and type of soil formed is ________.

19. If a soil has had most of its ions leached from the A horizon, but maintains a high organic content, it is classified as a ________.

20. A concentration of calcium carbonate in the B-horizon is called ________ and is characteristic of _______ soils.

21. The concentration through leaching of aluminum hydroxides is characteristic of _______ soils and are a source of aluminum ore called ________.

22. The more time a rock body has been exposed to soil formation, the ________ the soil profile.

23. ________ soils cause millions of dollars in damage to structures built on them because of their reaction to repetitive wetting and drying.

24. The two most significant factors in soil erosion are ________ and ________.

25. Weathering can lead to the formation of ore bodies by the selective removal of soluble substances to produce a ________ concentration.

MULTIPLE CHOICE

1. Talus cones are most abundant in areas
   a. where abundant rainfall is available to dissolve limestone.
   b. in tropical areas where thick soil and abundant vegetation occurs.
   c. in high mountains that have many days of sub-freezing temperatures.
   d. none of the above

2. Exfoliation domes form from
   a. frost wedging.
   b. pressure release.
   c. hydrolysis.
   d. root wedging.

3. Plants may play an important role in
   a. mechanical weathering.
   b. chemical weathering.
   c. both a and b
   d. neither a nor b

4. Which of the following is NOT a mechanical weathering process?
   a. frost wedging
   b. pressure release
   c. hydrolysis
   d. root wedging

5. An important acid in the weathering of limestone is
   a. sulfuric acid.
   b. hydrochloric acid.
   c. acetic acid.
   d. carbonic acid.

6. Limestone and marble chemically weather by
   a. oxidation.
   b. hydrolysis.
   c. solution.
   d. all of these

7. Oxidation has the greatest effect on
   a. feldspars.
   b. ferromagnesian silicates.
   c. nonferromagnesian silicates.
   d. carbonates.

8. Feldspars weather to clay by
   a. solution.
   b. hydrolysis.
   c. oxidation.
   d. none of these

9. The asymmetry of the water molecule makes it an important agent of
   a. exfoliation.
   b. frost heaving.
   c. solution.
   d. oxidation.

10. Which of the following rock types most commonly displays spheroidal weathering?
    a. limestone
    b. granite
    c. rock salt
    d. all equally

11. Chemical weathering is most intense in a
    a. hot, dry climate.
    b. cold, wet climate.
    c. hot, wet climate.
    d. cold, dry climate.

12. Which of the following is an example of regolith?
    a. soil
    b. sand dunes
    c. volcanic ash
    d. all of the above

13. The C-horizon of a soil has more ____ than the horizons above it.
    a. clay minerals
    b. organic matter
    c. rock fragments of the parent rock
    d. none of these

14. The A horizon of a soil is the
    a. zone of accumulation.
    b. zone of leaching.
    c. zone of caliche.
    d. zone of calcification.

15. The O horizon of a soil is rich in
    a. clay minerals.
    b. organic matter.
    c. rock fragments of the parent rock.
    d. none of these

16. The factor most important in determining soil type and depth is
    a. parent material.
    b. plants.
    c. climate.
    d. they are all equal

17. Pedocals are
    a. rich in calcium.
    b. found in the west.
    c. formed mostly in arid areas.
    d. all of the above

18. Caliche is an accumulation of
    a. calcium carbonate in the A horizon.
    b. calcium carbonate in the B-horizon.
    c. aluminum in the B-horizon.
    d. none of these

19. Laterite soils
    a. are red in color.
    b. are formed in tropical climates.
    c. are not very fertile.
    d. all of the above

20. Which is not a process of soil degradation?
    a. salinization
    b. residual concentration
    c. erosion
    d. compaction

21. Bauxite forms from
    a. the residual concentration of aluminum in the soil.
    b. oxidation reactions with descending metal rich fluids just above the water table.
    c. hydrothermal alteration of granitic rocks.
    d. all of these

22. Minerals formed by oxidation include
    a. clays.
    b. calcite.
    c. hematite.
    d. pyroxene.

23. Detrital sedimentary rocks
    a. are made of pieces of pre-existing rocks.
    b. have a crystalline texture.
    c. form by evaporation.
    d. all of these answers

24. Nature can break rocks apart using
    a. growing ice crystals.
    b. growing salt crystals.
    c. release of pressure by removal of overburden.
    d. all of these answers

25. The key difference between regolith and soil is
    a. thickness.
    b. regolith includes weathering residue.
    c. soil includes organic material.
    d. soil develops faster than regolith.

TRUE OR FALSE

____1. Erosion is the physical and chemical breakdown of rocks.

____2. Frost heaving is due to the expansion of ice as water freezes in unconsolidated sediment and soil.

____3. Exfoliation may produce large rounded domes of rock in mountainous areas.

____4. The growth of salt minerals in the cracks of rock is an important chemical weathering process.

____5. Plants and animals can bring about mechanical weathering.

____6. Water is an essential ingredient in chemical weathering.

____7. The red color in many soils is the result of the oxidation of aluminum.

____8. Most clay minerals result from the hydrolysis of calcite.

____9. As the particle size decreases, the rate of chemical weathering increases.

____10. Some rocks are more resistant to chemical weathering than others.

____11. Quartz is the most resistant of the common rock forming minerals.

____12. Spheroidal weathering occurs because the flat faces of a rectangular block of rock weather faster than the corners.

____13. Soil is a type of regolith that can support plant life.

____14. To be fertile and productive a soil must contain at least 50% humus.

____15. Residual soils form on stream sediments.

____16. The B-horizon is a humus-rich zone.

____17. Caliche is a soil rich in iron and aluminum.

____18. Bacteria, insects and worms can destroy a soil's fertility and, therefore, should be eliminated.

____19. The longer an outcrop is exposed to soil formation, the thicker the resulting soil profile.

____20. Expansive soils should be kept wet to minimize the damage to a building's foundation.

____21. Rill erosion occurs when rainwater flows down a slope in a small channel called a rill.

____22. Erosion becomes a problem when it is rapid enough to remove the A horizon exposing the B horizon below it.

____23. Such methods as crop rotation, terracing, contour plowing, and no-till planting were found to be a major cause of soil degradation and therefore are no longer widely practiced.

____24. Bauxite is an ore of iron.

____25. Gossans form at the water table.

DRAWINGS AND FIGURES

1. Identify and explain the origin of the weathering features shown in the photographs below.

(a)

(b)

2. Draw a typical soil profile.

Chapter 7

Sediment and Sedimentary Rocks

CHAPTER OBJECTIVES

By the end of this chapter you should be able to:

1. Differentiate between detrital and chemical sediment.
2. Classify detrital sedimentary particles.
3. Describe what happens to detrital sedimentary particles during transport.
4. List several examples of sedimentary environments.
5. Describe the roles of compaction and cementation in of various sedimentary rocks.
6. Classify detrital sedimentary rocks and list most abundant constituents in each type.
7. Give the names and mineral compositions of the most abundant chemical and biochemical sedimentary rocks.
8. Explain what sedimentary facies are and how they become layers.
9. Know what marine transgressions and regressions are, their effects on the geologic record, and some of the possible causes for them.
10. Describe and recognize several sedimentary structures, know how they form, and what they say about the depositional environment.
11. Describe several ways in which fossils are preserved.
12. Be familiar with the important natural resources associated with sedimentary rocks.

A USEFUL ANALOGY

One of the harder concepts in geology for many people is how a depositional environment like a beach, which deposits in a very narrow zone, can become a layer that may cover hundreds of square miles. Related to this is the problem of getting facies that are side by side to ultimately form vertically stacked layers. After all, there are no beaches hundreds of miles wide. The key is understanding the element of time and the concept of facies migration.

Imagine three workmen working in a house. They are in a big hurry to finish putting a carpet down with a plastic stain barrier on top of the pad. So all three are working at once. The first workman is rolling out the carpet pad. Right on his heels is the chap rolling out the plastic stain barrier. The third is rolling out the carpet itself on top of the plastic. At any point in time if you say "stop", each of the three will be in the act of rolling out his respective material. They will be doing this at only a single spot and they will be beside one another in space. This is like looking at a depositional system at any given point in time. Each workman is a facies depositing a different material. However, if you go back to where all three have been you will see a three part layering with a carpet pad on the bottom followed by the plastic and then the carpet... sedimentary layers!

KEY TERMS

After reading this chapter, you should be familiar with the following terms:

sediment
sedimentary rock
sediment transport
detrital sediment
gravel
sand
silt
clay
chemical sediment
abrasion
rounding
sorting
depositional environment
lithification
pore space
compaction
cementation
detrital sedimentary rock
clastic texture
conglomerate
breccia
rubble
sandstone
quartz sandstone
arkose
mudrock
siltstone
mudstone
claystone
shale
chemical sedimentary rock
crystalline texture
biochemical sedimentary rocks
limestone
dolostone
carbonate rocks
travertine
oolitic limestone
coquina
chalk
evaporites
rock salt
rock gypsum
chert
flint
jasper
nodules
coal
peat
lignite
bituminous
anthracite
sedimentary facies
marine transgression
marine regression
sedimentary structures
strata
sedimentary bed
bedding plane
cross-bedding
paleocurrents
graded bedding
turbidity current
ripple marks
mud cracks
fossil
body fossil
trace fossil
petrification
mold
cast
formation
silica sand
placer deposit
hydrocarbons
source rock
reservoir rock
permeabiltiy
stratigraphic trap
structural trap
salt domes
oil shale
kerogen
tar sands
coke
carnotite
uraninite
banded iron formations

CHAPTER CONCEPT QUESTIONS

1. Describe what happens to detrital particles during the transportation process.

2. Compare and contrast the lithification of sand or gravel with that of mud.

3. A typical granite has about 60% feldspar and 15% quartz. However, sandstone derived from granite typically has over 60% quartz and 25% feldspar. Explain this reversal in abundance.

4. How do chemical sedimentary rocks differ in origin from detrital sedimentary rocks? How is this difference reflected in their texture?

5. Describe two ways in which limestone form. As an example contrast the origin of oolitic limestone with that of coquina.

6. Explain the origin of evaporites and give two examples.

7. How does chert form?

8. How do coal, lignite, and peat differ from other chemical sedimentary rocks?

9. What are sedimentary facies? How do they ultimately form layered rocks over time?

10. Describe two possible causes of marine transgressions.

11. What are sedimentary structures, and how are they used? Give several examples of such structures and explain what information each provides about the origin of the rock in which you find them.

12. What is the difference between trace fossils and body fossils?

13. Describe several ways in which an organism can be fossilized.

14. Describe some of the characteristics of sedimentary rocks studied by geologists in identifying the depositional environment of sedimentary rocks.

15. Explain the origin of oil and natural gas. How do they accumulate in economically viable deposits?

16. What are oil shales and tar sands? How might they help solve our future energy problems? What problems are associated with their exploitation.

COMPLETE THE FOLLOWING TABLES

DETRITAL SEDIMENTARY ROCKS			
Clast size	**Sediment Name**	**Rock Name**	
> 2 mm		(rounded) (angular)	
1/16-2 mm			
1/256-1/16 mm			(mixture)
<1/256 mm			

CHEMICAL and BIOCHEMICAL SEDIMENTARY ROCKS	
Composition	**Rock Name**
Calcite	
Dolomite	
Gypsum	
Halite	
Quartz (microscopic)	
Carbon	

COMPLETION QUESTIONS

1. Sediment forms from the ________ of pre-existing rocks.

2. Sediment composed of the particles of disintegrated pre-existing rock is called ________ sediment.

3. Sedimentary particles are transported by wind, ________, and ice to a site of deposition.

4. While sediment is being transported, it is reduced in size and made smoother by ________.

5. During transport, sediment is segregated by size in a process called ________.

6. Transportation ends when sediment is deposited in an area called a ________.

7. Once sediment is deposited, it is buried and converted to sedimentary rock in a process called ________.

8. In converting mud to shale, ________ alone is generally sufficient as a lithification process, but in converting sand to sandstone and gravel to conglomerate ________ is also important.

9. Minute amounts of the mineral ________ in cement can give a red color to sandstone.

10. The fragments composing sedimentary rocks are called ________ and the texture exhibited by these rocks is, therefore, called a ________ texture.

11. Conglomerate and breccia are composed of ________-sized clasts, which are larger than ________ millimeters in diameter.

12. Compared to breccia, conglomerates display a greater degree of ________.

13. The most abundant mineral found as clasts in sandstone is _______.

14. The finest-grained detrital sedimentary rock is _______.

15. If mudstone or claystone displays ______ it is called shale.

16. A biochemical limestone composed entirely of broken shells is called ________.

17. ________ are small spherical grains of inorganically precipitated calcite.

18. Halite and gypsum precipitate when ions in solution are concentrated by ________ of water.

19. Chert is a chemical sedimentary rock composed of ________.

20. ________ is composed of the compressed and altered remains of plants that were deposited in waters deficient in ________.

21. Coal grades from peat (the lowest grade) to ________, to ________, to ________ (the highest grade).

22. During a marine ________, the sea advances on to a continent, and during a ________ the sea moves off of a continent.

23. Features that form as sediments are being deposited are called sedimentary _________.

24. Geologists use sedimentary structures to determine the environment of ________.

25. _________ can be used to determine the direction of flow of ancient currents.

26. A shell-shaped cavity caused by the dissolution of that shell is called a _________, and if this is filled-in to make a replica of the fossil, it is called a ________.

27. _______, _______ and _______ are energy resources that are found in sedimentary rocks.

28. Hydrocarbons form in a _________ rock, and accumulate in a ________ rock.

29. Over 50% of the world's proven reserves of petroleum are in the ________ region.

30. _________, an organic substance that can be extracted from oil shale and may someday prove a viable source of fossil fuels.

31. Banded iron formations, important sources of iron ore, are almost all _________ in age.

MULTIPLE CHOICE

1. Sedimentary rocks formed from particles of older rock broken up by weathering are called
   a. chemical sedimentary rocks.
   b. evaporites.
   c. detrital sedimentary rocks.
   d. biochemical sedimentary rocks.

2. Chemical sedimentary rocks form from
   a. broken fragments of pre-existing rock.
   b. precipitation from solution.
   c. wind blown sediments.
   d. all of the above

3. Sand size particles are
   a. 2 to 1/16 mm in diameter.
   b. greater than 2 mm in diameter.
   c. less than 1/15 mm in diameter.
   d. 1/16 to 1/256 mm in diameter.

4. Sediments are transformed into sedimentary rocks by
   a. weathering and abrasion.
   b. compaction and cementation.
   c. sorting and abrasion.
   d. none of these

5. The most common cementing minerals are
   a. calcite and quartz.
   b. feldspar and quartz.
   c. hematite and quartz.
   d. calcite and hematite.

6. Conglomerates are composed of
   a. angular grains larger than 2 mm in diameter.
   b. grains 2 to 1/16 mm in diameter.
   c. rounded grains larger than 2 mm in diameter.
   d. none of these

7. An arkose is
   a. sandstone composed of 90% quartz.
   b. a lithified gravel with angular clasts.
   c. a fissile mudrock.
   d. a sandstone with at least 25% feldspar.

8. Shales are
   a. fissile.
   b. composed of particles <1/16 mm in diameter.
   c. the most abundant sedimentary rock.
   d. all of these

9. Limestone is composed of
   a. clay minerals.
   b. calcium carbonate.
   c. silica.
   d. none of these

10. Chalk is a type of
   a. limestone composed of microscopic shell fragments.
   b. limestone composed of oolites.
   c. sandstone composed of substantial amounts of feldspar.
   d. mudrock having fissility.

11. Chert is composed of
   a. halite.
   b. calcite.
   c. quartz.
   d. gypsum.

12. Coal forms from
   a. microscopic plankton.
   b. silica precipitated from seawater.
   c. calcite precipitated from seawater.
   d. none of these

13. The form of coal with the highest carbon contents is
   a. anthracite.
   b. peat.
   c. lignite.
   d. bituminous.

14. Evaporites are
   a. inorganically precipitated chemical sedimentary rocks.
   b. biochemical sedimentary rocks.
   c. detrital sedimentary rocks.
   d. none of the above

15. A transgressive sea
   a. produces no facies.
   b. moves off of a continent.
   c. advances onto a continent.
   d. produces layers with shale underlying sandstone.

16. Periods of rapid episodes of seafloor spreading at divergent margins likely result in
    a. long periods of stable sea level at the margins of continents.
    b. periods of marine transgression.
    c. periods of marine regression.
    d. the formation of glaciers.

17. The most important use of sedimentary structures is to determine
    a. the environment of deposition.
    b. the age of a sedimentary rock.
    c. what economically important minerals might be present.
    d. what forms of life were present when the sediment was deposited.

18. Ancient current directions can be determined from
    a. mud cracks.
    b. current ripple marks and cross-bedding.
    c. graded bedding.
    d. trace fossils.

19. Fossils can be
    a. composed of the original bone and shell material.
    b. composed of bone and shells with their original composition replaced by a different mineral.
    c. composed of carbon films.
    d. all of these

20. In determining the depositional environment of a sedimentary rock, which of the following are studied?
    a. sedimentary structures
    b. texture and composition
    c. fossils
    d. all of the above

21. A lens of permeable sandstone within an impermeable shale may be a
    a. source rock for hydrocarbons.
    b. a structural trap for hydrocarbons.
    c. a stratigraphic trap for hydrocarbons.
    d. all of the above

22. The richest uranium deposits are associated with
    a. evaporite deposits.
    b. banded iron formations.
    c. plant fossils in non-marine sedimentary rocks.
    d. fish fossils in deep marine sedimentary rocks.

23. Banded iron formations contain alternating layers of magnetite and which chemical sedimentary rock?
    a. limestone
    b. rock salt
    c. rock gypsum
    d. chert

24. Graded bedding
    a. forms in underwater landslides.
    b. are layers that increase in thickness laterally.
    c. typically form in glacial deposits
    d. are layers deposited as wind storms die down.

25. Lithification
    a. may include cementation and compaction.
    b. applies only to chemical sediments.
    c. results in net increase in volume.
    d. occurs only on convergent boundaries since it requires tremendous pressure.

TRUE OR FALSE

____1. Sedimentary rocks make up only 5% of the crust but are the most common rock type at the surface.

____2. Detrital sediments form by precipitation from solution.

____3. Sand size sediments are larger than 2 mm in diameter.

____4. Sorting refers to the uniformity of the grain sizes present.

____5. Compaction alone is generally enough to lithify mud into shale.

____6. Chemical sediments are primarily lithified by cementation.

____7. Large clasts are readily rounded by water and so conglomerates are more abundant than sedimentary breccias.

____8. Mudrocks may have both silt and clay-sized material.

____9. Calcite is a common cementing mineral in sandstone.

____10. Oolitic limestone is composed almost entirely of broken shell material.

____11. Dolomite generally precipitates directly out of seawater.

____12. The mineral found in rock salt is gypsum.

____13. A likely source for silica in chert is the dissolved shells of microorganisms.

____14. Coal forms as land plant debris accumulates in swampy water rich in oxygen.

____15. Bituminous coal has a higher carbon content than lignite.

____16. A facies is characterized by a distinctive set of physical, chemical, and biological attributes.

____17. Deep sea shales are likely to be characterized by the presence of mud cracks.

____18. Graded bedding shows an upward decrease in grain size.

____19. Most banded iron formations formed over 2 billion years ago.

____20. When the original shell of a fossil dissolves it leaves a trace fossil.

____21. Permeability refers to the capacity of a rock or sediment to transmit fluid.

____22. Pure quartz sandstone is often mined for making glass.

____23. Low permeability shale tends to make the best reservoir rocks.

____24. The U.S. has most of the world's known oil shales.

___25. The richest uranium ores in the United States are found in the Colorado Plateau.

DRAWINGS AND FIGURES

1. Identify the sedimentary structure in the photograph below. Indicate with an arrow the direction of flow recorded by this structure.

2. Label the following depositional environments in the block diagram below.

alluvial fan	lagoon
barrier island	lake
beach	organic reef
continental shelf	sand dunes
deep marine environment	stream environment
delta	submarine fan
glacial environment	tidal flat

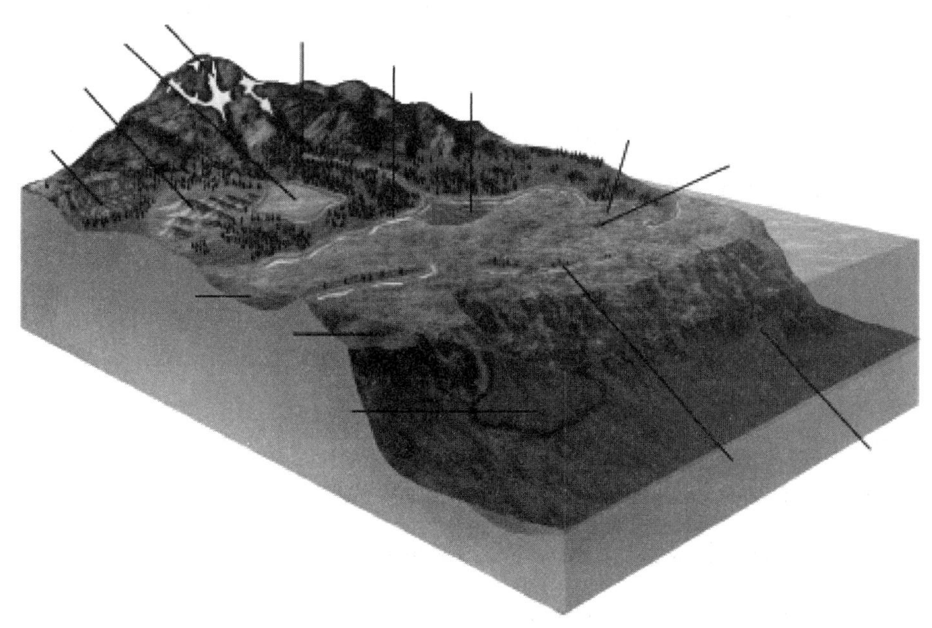

3. The cross-section illustrates the distribution of facies adjacent to a shoreline. Sketch the vertical sequence of sedimentary rocks resulting from (a) a marine transgression and (b) a marine regression.

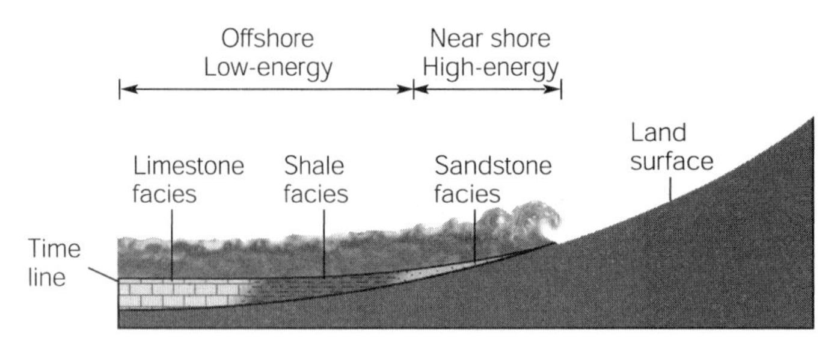

4. Identify the sedimentary structures in the following photographs.

a.

b.

Chapter 8
Metamorphism and Metamorphic Rocks

CHAPTER OBJECTIVES

By the end of this chapter you should be able to:

1. Explain why, and under what circumstances rocks metamorphose.
2. Locate the major shield areas of Earth.
3. Discuss how heat, pressure and fluid activity influence metamorphism.
4. Differentiate between contact, regional and dynamic metamorphism.
5. List the index minerals of metamorphic zones in order of increasing intensity of metamorphism.
6. Define and discuss the origin of foliation.
7. Classify metamorphic rocks listing important minerals, metamorphic grade, and parent rock of each.
8. Discuss the concepts of metamorphic zones and facies.
9. Relate types of metamorphism and metamorphic facies to plate tectonic setting.
10. List several economically important metamorphic rocks and minerals.

USEFUL ANALOGIES

It is easy to understand the difference between lithostatic and directed pressure if you realize that lithostatic pressure is a confining pressure and, therefore, much like two other types of confining pressure that we experience: hydrostatic pressure and atmospheric pressure. When you dive to the bottom of a swimming pool you do not feel the weight of the water column on your back. Rather, you experience an increase in pressure equally from all sides. This is may be manifested by a pain in your sinuses as the water tries to change the volume of your head much the same way it did the Styrofoam cup in Fig. 8.7a. When you rise and descend in an airplane you experience a similar change in pressure. As rocks are buried they also experience a similar change in confining pressure, in this case lithostatic pressure. You experience differential pressure whenever you feel the weight of an object on top of you. Pressure in this case is not equal in all directions.

By pretending to be a mineral you can see how differential pressure controls growth direction while confining pressure does not. Jump in a swimming pool and curl up in a fetal position. You are a mineral just starting to nucleate. To grow, stretch to the full length of your body. In a pool it doesn't matter what orientation you stretch. It is equally easy to grow in any direction. Now come out of the pool and curl up in a fetal position on the floor. Have a friend gently put a heavy mattress on top of you (not too heavy!) You are now feeling the weight of the mattress and so you are experiencing differential pressure. Now try to grow by stretching to the full length of your body. While it is easy to grow in a horizontal orientation it is not so easy to grow vertically, against the directed pressure.

KEY TERMS

After reading this chapter, you should know the following terms.

metamorphic rock
shield
lithostatic pressure
recrystallization
differential pressure
fluid activity
contact metamorphism
aureoles
hydrothermal alteration
dynamic metamorphism
mylonite
regional metamorphism
index minerals
foliated texture
slate
phyllite
schist
schistosity
gneiss
amphibolite
migmatite
nonfoliated texture
marble
quartzite
greenstone
hornfels
anthracite
isograd
metamorphic zone
metamorphic facies
graphite

CHAPTER CONCEPT QUESTIONS

1. What information about the earth does the study of metamorphic rocks provide?

2. What are the major changes that occur in a rock during metamorphism and what factors bring about those changes?

3. What is the effect of heat on rocks and what are the sources of the heat?

4. Explain the difference between lithostatic and differential pressure.

5. What is the effect of chemically active fluids on rocks and what are the sources of those fluids?

6. What are the three types of metamorphism and what are their major differences?

7. How does contact metamorphism take place? What are the major factors controlling the products of contact metamorphism?

8. What is dynamic metamorphism and under what circumstances does it occur?

9. How does regional metamorphism take place?

10. List the index minerals of regional metamorphism in order of increasing metamorphic intensity.

11. What is foliation and how is it produced? Why don't contact metamorphic rocks generally display foliation? Why are marbles and quartzites, even when created by regional metamorphism, nonfoliated?

12. What are metamorphic zones and metamorphic facies? How are they similar and how different?

13. List at least five economically important resources created by metamorphism.

COMPLETE THE FOLLOWING TABLES

FOLIATED METAMORPHIC ROCKS			
Parent Rock(s)	**Metamorphic Grade**	**Characteristics**	**Metamorphic Rock**
mudrock, claystone volcanic ash	low	fine-grained, dull	
mudrock	low to medium	fine-grained, lustrous	
mudrock, impure carbonates, mafic igneous rock	low to high	distinct foliation, visible grains	
mudrock, sandstone, felsic igneous rock	high	light and dark minerals segregated into bands	
mafic igneous rock	medium to high	Hornblende dominates, dark, weak foliation.	

NONFOLIATED METAMORPHIC ROCKS			
Parent Rock(s)	**Metamorphic Grade**	**Characteristics**	**Metamorphic Rock**
limestone or dolostone	low to high	interlocking calcite and dolomite	
quartz sandstone	medium to high	interlocking quartz crystals	
mafic igneous rocks	low to high	fine grained, green color	
mudrock	low to medium	fine-grained, hard and dense	
coal	high	black, lustrous up to 98% carbon	

FOLIATED METAMORPHIC ROCKS			
Parent Rock(s)	**Metamorphic Grade**	**Characteristics**	**Metamorphic Rock**
	low	fine-grained, dull	Slate
	low to medium	fine-grained, lustrous	Phyllite
	low to high	distinct foliation, visible grains	Schist
	high	light and dark minerals segregated into bands	Gneiss
	medium to high	Hornblende dominates, dark, weak foliation.	Amphibolite

NONFOLIATED METAMORPHIC ROCKS			
Parent Rock(s)	**Metamorphic Grade**	**Characteristics**	**Metamorphic Rock**
	low to high	interlocking calcite and dolomite	Marble
	medium to high	interlocking quartz crystals	Quartzite
	low to high	fine grained, green color	Greenstone
	low to medium	fine-grained, hard and dense	Hornfels
	high	black, lustrous up to 98% carbon	Anthracite

COMPLETION QUESTIONS

1. Metamorphic rocks form from preexisting rocks when those rocks are subjected to conditions that change their ________ composition and/or ________.

2. Precambrian metamorphic rocks are exposed in vast areas of the continents called ________.

3. The agents of metamorphism are ________, ________, and chemically active ________.

4. Sources of heat include rising _______, and the ________ gradient.

5. ________ pressure is due to the weight of the overlying rocks and is applied equally in all directions.

6. In mountain belts, pressures from two directions often exceed pressures from other directions. These are called _________ pressures.

7. The presence of water ________ the speed of metamorphic chemical reactions.

8. Sources of water for metamorphic reactions include water trapped in the ________ of sedimentary rocks, volatile fluids from rising _______, and the release of water by the ________ of certain minerals.

9. The three major types of metamorphism are ________, ________, and ________ metamorphism.

10. _______ metamorphism is due to the "cooking" of country rock by magma.

11. Because temperature decreases away from a heat source, concentric zones of different mineral assemblages commonly surround intrusions. These altered areas are known as metamorphic _______.

12. _______ metamorphism is generally restricted to fault zones.

13. The burial of large areas of rock into realms of high temperatures and pressures results in _______ metamorphism.

14. Foliated metamorphic rocks derive their texture from subjection to high _______ pressures.

15. _______ is a very fine-grained, low-grade metamorphic rock, valuable because of its propensity to split along very flat "cleavage" planes.

16. _______ is composed of alternating bands of light colored quartz and feldspar and dark colored ferromagnesian minerals.

17. _______ appear to be mixtures of high grade metamorphic rocks and streaks of granitic igneous rocks.

18. Marbles are generally composed of the mineral _______ and/or dolomite and result from metamorphism of ________.

19. A very hard, strong metamorphic rock composed almost entirely of quartz is called _______.

20. Hornfels results from the _______ metamorphism of mudrock.

21. The highest grade coal, _______, is actually a metamorphic rock.

22. A group of metamorphic rocks that are characterized by particular mineral assemblages formed under the same broad temperature pressure conditions is called a metamorphic ________.

23. Lines of equal metamorphic intensity plotted on a map are called _______.

24. Mylonites form from pure __________ metamorphism along _______.

25. Sillimanite is an example of an ___________ mineral formed under _________temperature.

MULTIPLE CHOICE

1. Metamorphic rocks form from
   a. igneous rocks only.
   b. sedimentary rocks only.
   c. igneous and sedimentary rocks only.
   d. igneous, sedimentary, and metamorphic rocks.

2. Metamorphism can occur
   a. only after a preexisting rock melts and begins to cool.
   b. only in the solid state, before a rock melts.
   c. either a or b
   d. none of these

3. Metamorphism can cause a change in the
   a. texture of the pre-existing rock.
   b. mineral composition of the pre-existing rock.
   c. a and b
   d. none of these

4. With increasing depth
   a. temperature generally increases.
   b. lithostatic pressure increases.
   c. both a and b
   d. neither a nor b

5. The presence of water
   a. increases the rate of chemical reactions.
   b. reduces the rate of chemical reactions.
   c. has no effect on the rate of chemical reactions.
   d. prevents metamorphism from occurring.

6. Which factor is least important in contact metamorphism?
   a. heat
   b. high differential pressure
   c. presence of fluids
   d. all of the above are equally important

7. Contact metamorphism dominates
   a. in fault zones.
   b. at shallow depths around intrusions.
   c. in the cores of mountain belts.
   d. none of the above

8. Which of the following is most important in dynamic metamorphism?
   a. high lithostatic pressure
   b. high differential pressure
   c. presence of fluids
   d. heat

9. Dynamic metamorphism dominates
   a. in fault zones.
   b. at shallow depths around intrusions.
   c. in the cores of mountain belts.
   d. none of the above

10. Regional metamorphism dominates
    a. in fault zones.
    b. at shallow depths around intrusions.
    c. in the cores of mountain belts.
    d. none of the above

11. Regional metamorphism usually involves an increase in
    a. heat.
    b. lithostatic pressure.
    c. differential pressure.
    d. all of the above

12. Which of the minerals listed below indicates the highest grade of metamorphism?
    a. chlorite
    b. garnet
    c. sillimanite
    d. biotite

13. Which of the following lists contains all foliated rocks?
    a. schist, gneiss, quartzite
    b. slate, schist, gneiss
    c. schist, gneiss, marble
    d. none of the above

14. The parent rock for marble is
    a. sandstone.
    b. mudrock.
    c. limestone.
    d. any of the above

15. Which of the following is not metamorphic in origin?
    a. shale
    b. slate
    c. schist
    d. serpentine

16. Schist and gneiss form by
    a. contact metamorphism.
    b. regional metamorphism.
    c. dynamic metamorphism.
    d. hydrothermal alteration.

17. Amphibolite forms by metamorphism of rocks initially rich in
    a. ferromagnesian minerals.
    b. quartz and feldspar.
    c. clay minerals.
    d. calcite.

18. Migmatite appears to be a mixture of
    a. granite and gneiss.
    b. basalt and slate.
    c. sandstone and schist.
    d. shale and phyllite.

19. Subduction zones may be recognized by the presence of the _______ facies.
    a. zeolite
    b. amphibolite
    c. greenschist
    d. blueschist

20. The most widespread regional metamorphism can be found
    a. at divergent margins.
    b. at convergent margins.
    c. at transform boundaries.
    d. over hot spots in the middle of plates.

21. The primary source of fluids during metamorphism does NOT include
    a. buried gypsum and clay minerals.
    b. underground lakes.
    c. sedimentary rocks.
    d. magmatic volatiles.

22. Study of metamorphic rocks is important because
    a. metamorphic rocks form much of the continental crust.
    b. metamorphic rocks may contain gemstones.
    c. some metamorphic minerals have industrial uses.
    d. all of these answers

23. Why do foliated metamorphic rocks form at convergent margins?
    a. The collision creates high temperatures and pressures.
    b. The temperatures are too low to form hornfels.
    c. The temperatures are too high to form nonfoliated metamorphic rocks.
    d. Fluid activity increases the rate of chemical reactions.

24. The baked zone under a thick lava flow is an example of
    a. regional metamorphism.
    b. dynamic metamorphism.
    c. contact metamorphism.
    d. lithosphere metamorphism.

25. Heat for metamorphism comes from
    a. the Sun.
    b. deep burial and magma intrusions.
    c. chemical reactions.
    d. pressure on fluids.

TRUE OR FALSE

____1. Time is not an important factor in metamorphic reactions.

____2. Metamorphic and sedimentary rocks form the crystalline basement rocks found in the shields of the continents.

____3. An increase in temperature can increase the speed of metamorphic reactions.

____4. Metamorphic grade decreases with increasing distance from the intrusion.

____5. Increases in lithostatic pressure impart a change in both volume and shape of a mineral.

____6. Fluids slow down the rate of metamorphic reactions.

____7. Contact metamorphism occurs when the rocks are in contact with a fault.

____8. Contact metamorphism can cause new minerals to form.

____9. Metamorphic aureoles adjacent to large plutons are generally wider than those that are next to small plutons.

____10. Hot, watery solutions from a crystallizing magma produce new metamorphic minerals by hydrothermal alteration of country rock.

____11. A mylonite is produced by hydrothermal alteration of country rock.

____12. The mineral composition of a metamorphic rock is a poor indicator of the degree or intensity of the metamorphism.

____13. Foliation results from an increase in lithostatic pressure.

____14. Schist is characterized by having alternating bands of ferromagnesian and nonferromagnesian minerals.

____15. Phyllite generally indicates a higher grade of metamorphism than a gneiss.

____16. Mafic igneous rocks may metamorphose into amphibolite or greenstone.

____17. Anthracite is a metamorphic form of coal.

___18. Hornfels is produced when shale is subjected to high temperatures but low differential pressure.

___19. A metamorphic facies is named after its most characteristic rock or mineral.

___20. Sediments deposited in a trench are sheltered from metamorphism.

___21. Divergent plate boundaries are likely places for contact metamorphism to occur.

___22. Metamorphism is not found at divergent boundaries.

___23. Blueschist facies rocks form under higher temperatures than amphibolite.

___24. Foliation indicates high temperatures.

___25. Metamorphism can obliterate evidence of the original rock.

FIGURES AND DIAGRAMS

1. In the diagram below indicate where you would expect (a) contact (high temperature, low pressure), regional (high pressure, high temperature), and blueschist (high-pressure, low temperature) metamorphism to occur. Be able to defend your choices.

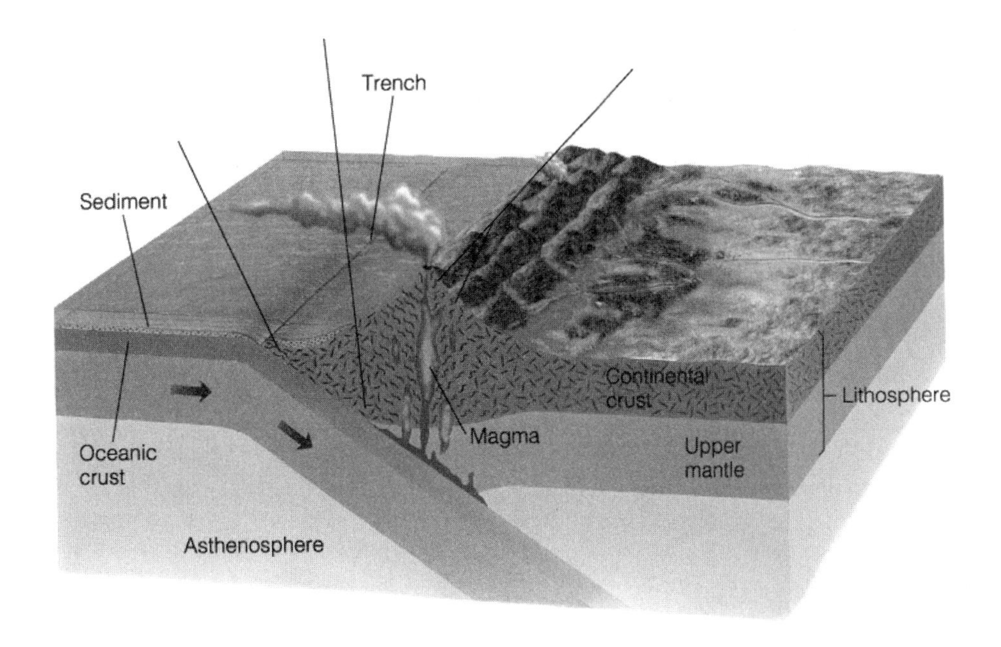

2. Using the diagram below answer following questions.
   a. Which facies would you expect to find in a schist metamorphosed from sediments deposited in a trench where high pressures but low temperatures exist?
   b. As you progress away from a shallow (pressure = 1 kb) pluton which has a temperature of 700 degrees C, what progression of metamorphic facies would you expect to encounter?
   c. As you progress away from the core of a mountain belt, where the most intense regional metamorphism occurred, toward the unmetamorphosed country rock, what progression of metamorphic facies would you expect to encounter?

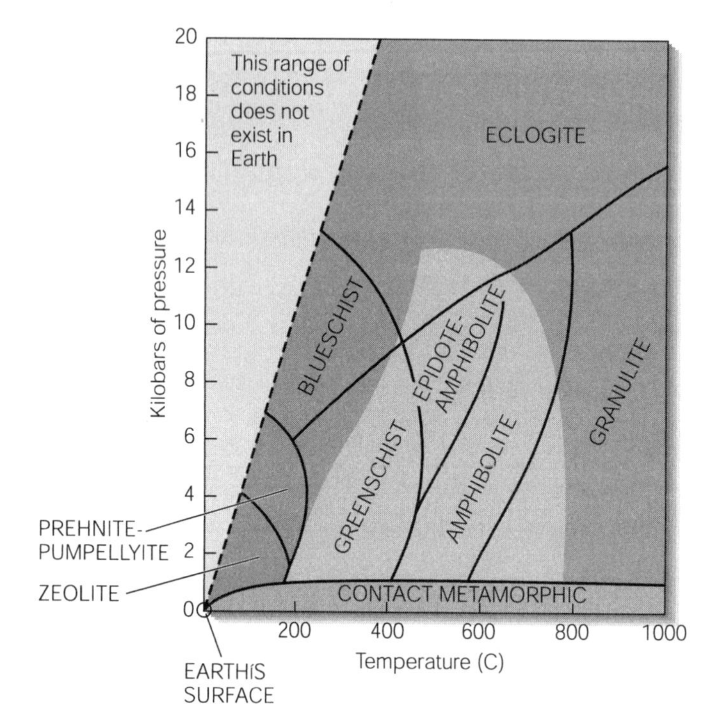

Chapter 9

Geologic Time: Concepts and Principles

CHAPTER OBJECTIVES

By the end of this chapter you should be able to:

1. Distinguish between relative dating and absolute dating.
2. Summarize the contributions to our view of geologic time of such notable scientists as James Hutton, Nicholas Steno, Lord Kelvin, Charles Lyell, Georges Louis de Buffon, and William "Strata" Smith.
3. List, define and apply those principles which are used in relative dating noting those that are attributed to Nicholas Steno.
4. Explain what an unconformity is, what it implies and classify the different types of unconformities.
5. Discuss the goals and techniques of correlation.
6. Recall the different parts of an atom.
7. Differentiate between the three types of radioactive decay and give examples of isotope parent-daughter pairs that exhibit each type.
8. Perform simple calculations of rock ages using half-lives.
9. Discuss the assumptions behind, and sources of uncertainty in radioactive dating.
10. Explain the principle behind fission track dating.
11. Explain technique and limitations of Carbon-14 dating.
12. Explain the principles behind dating using tree rings.
13. Reproduce the geologic time scale to the extent that your instructor requires.

USEFUL ANALOGIES

The vastness of geologic time is one of the hardest thing for mortals to grasp. Numbers like 4.5 billion are so unfamiliar to most people that to say that the Earth is that old is likely to elicit nothing but a blank stare. Over the years a number of techniques have been used to try to put the age of the Earth and the relative timing of events in some sort of perspective.

One popular way is to equate the age of the earth to a calendar year. The Earth was created at just an instant after midnight on January 1 and the present day is at midnight the following December 31. The oldest fossil life appears sometime before sunrise on March 25. However, abundant fossil life, with shells, that we all would recognize as fossils don't appear until between 6:00 and 6:30 PM November 15! The first land plants sprout up about 5:30 PM on Nov. 26 and the first amphibians crawl onto land just before midnight on Dec. 1. The first reptiles show up around lunchtime on Dec. 6 but the dinosaurs don't appear until 3:00 AM on Dec. 12. Mammals were a couple of days later, arriving on the scene about 345 on the 14th. Birds came along about 7:30 in the morning Dec. 20. The dinosaurs ruled the Earth until something killed them off mid-afternoon the day after Christmas. *Australopithecus* , a very early hominid finally shows up about 3:30 in the afternoon on Dec. 31 and genus *Homo,* our very own genus, makes an appearance at 8:00 PM. *Homo sapiens* finally settles in about 12 seconds before midnight!

Another way is to set up a time line. You can make it 4.5 miles long so that the scale is 1 mile to a billion years. Life, then, becomes abundant over the last .57 miles of your line. Dinosaurs show up .25 miles from the end and die off about 77.5 feet from the end. *Australopithecus* arrives 5.25 feet from the end and *Homo sapiens* spans the last inch and a half. By the way your life is less than 3/10000 of an inch long. Feeling a bit insignificant?

Combining relative and absolute dating techniques is not limited to geology. We really do it all the time. Imagine going to meet a friend at a restaurant. On the way you notice that it starts to rain at 5:30. When you arrive your friend's car is already there. What can you say about how long he or she has been waiting? Answer: Look under the car. If it is wet you know the your friend arrived after the rain and has not been waiting all that long!

KEY TERMS

After reading this chapter you should know the following terms.

relative dating
relative geologic time scale
absolute dating
principle of uniformitarianism
principle of superposition
principle of original horizontality
principle of lateral continuity
principle of cross-cutting relationship
principle of inclusions
principle of fossil succession
unconformity
hiatus
disconformity
angular unconformity
nonconformity
correlation
key bed
guide fossil or index fossil
concurrent range zone
well cuttings
well log
radioactive decay
alpha decay
beta decay
electron capture
half-life
parent element
daughter element
mass spectrometer
radon
fission track dating
carbon 14 dating
tree-ring dating
geologic time scale

CHAPTER CONCEPT QUESTIONS

1. What is the difference between relative time and absolute time?

2. Explain the principle of uniformitarianism. How did it revise our beliefs about the age of the Earth?

3. How did Kelvin calculate the age of Earth? What were the problems with his theory?

4. Explain, with the help of diagrams if necessary, each of Steno's principles.

5. Explain, with the help of diagrams if necessary, the principle of crosscutting relationships and the principle of inclusion. How do they help in relative dating of igneous and sedimentary rocks?

6. Explain the principle of fossil succession. How has it proven valuable in correlation?

7. What is an unconformity? Summarize characteristics of the three major types.

8. List the desirable characteristics of an guide fossil. How can fossils that are not guide fossils be used in correlation?

9. Describe three techniques used to correlate rocks in the subsurface.

10. Explain the differences between alpha decay, beta decay, and electron capture.

11. Why are sedimentary rocks usually not amenable to dating by radiometric techniques?

12. What are possible sources of error in dating rocks radiometrically?

13. Explain the principles behind dating by fission track method.

14. How does Carbon-14 form? Explain the principle behind its use in absolute dating.

15. How are tree rings used in absolute dating?

COMPLETION QUESTIONS

1. ________ dating places geologic events in sequential order without regard to age in years.

2. ________ dating gives an age of a sample expressed in years before present.

3. Currently our best estimate for the age of Earth is _______ years.

4. The principle of ________ asserts that geologic processes happening today have operated throughout geologic time.

5. The principle of superposition states that in undisturbed beds, the oldest beds are on the ________.

6. The principle of original horizontality states that ________ is deposited in horizontal layers.

7. If inclusions of one rock are found within another, the rock containing the inclusions is _______ than the rock from which the inclusions were derived.

8. If a fault cuts across a sandstone layer, the sandstone is ________ in age than the fault.

9. Gaps in the geologic record of an area are called ________.

10. Unconformities with flat lying beds overlying tilted beds are called ________ unconformities.

11. Nonconformities occur when beds of________ rocks overlie ________ or ________ rocks.

12. Disconformities are often difficult to identify and may depend on the study of the ________ in the sequence to recognize them.

13. Correlation is sometimes accomplished by using layers of distinctive composition called ________.

14. Guide fossils have a ________ geographic distribution and lived over a ________ time range.

15. If one does not have guide fossils to use, one might be able to correlate strata using several longer-lived species by employing __________.

16. Radioactive dates are based on the ratio of unstable ________ elements to their stable products called ________ elements.

17. In alpha decay a particle composed of two ________ and ________ neutrons are emitted from the nucleus.

18. In beta radiation a ________ in the nucleus changes to a ________ by emitting an ________.

18. The ________ of an element is the time it takes for half of the parent element to decay.

20. The amount of parent and daughter atoms present is determined by a sophisticated instrument called a ________ .

21. Carbon-14 dates are useful if the sample is less than ______ years old.

22. Carbon-14 decays to ________ by ________ decay and is not replenished once an organism dies.

23. Geologically recent events are sometimes datable by correlating the _______ of trees.

24. The geologic time scale combines information from both ____________ and ___________ dating.

25. The most useful rocks for relative dating are __________ rocks since they form in ________ over large areas and are the most likely to contain evidence of past life called ________.

MULTIPLE CHOICE

1. Absolute dates are based on
   a. fossils.
   b. educated estimates.
   c. radioactive decay.
   d. crosscutting relationships.

2. The method that gave an acceptable date for the age of the Earth was
   a. the cooling rate of the Earth.
   b. the rate of sedimentation.
   c. the salinity of the oceans.
   d. none of the above

3. The principle of uniformitarianism required that the age of the Earth
   a. must be much older than previously thought.
   b. must be much younger than previously thought.
   c. could only be deciphered from studying the salinity of the ocean.
   d. could be deciphered from radioactive materials.

4. In undisturbed sedimentary deposits, the oldest beds
   a. are usually lava flows.
   b. are on the top.
   c. are on the bottom.
   d. could be b or c

5. Which of the following principles is NOT attributed to Steno?
   a. original horizontality
   b. uniformitarianism
   c. lateral continuity
   d. All of the above ARE attributed to Steno.

6. The Earth is approximately
   a. 4.6 billion years old.
   b. 4.6 million years old.
   c. 46 million years old.
   d. 6000 years old.

7. The principle of lateral continuity states that
   a. beds end abruptly.
   b. beds gradually thin and pinch out.
   c. the oldest beds are on the bottom.
   d. that beds are deposited as flat lying units.

8. The principle of fossil succession states that
   a. fossils of a particular organism occur at different times in different locations.
   b. the sequence of fossils is the same all over and, therefore, is predictable.
   c. fossils cannot be relied upon in correlation.
   d. none of the above

9. Unconformities represent
   a. a gap in the rock record.
   b. a period of erosion or non-deposition.
   c. neither a or b
   d. both a and b

10. A disconformity occurs where
    a. the beds are parallel above and below the unconformable contact.
    b. the beds above and below the unconformable contact are angles to one another.
    c. sedimentary rocks overlie massive crystalline rocks.
    d. all of the above

11. A nonconformity may closely resemble an intrusive contact, particularly in concordant plutons. Which principle may help in making the distinction?
    a. lateral continuity
    b. uniformitarianism
    c. inclusions
    d. fossil succession

12. Rocks in different areas may be correlated by having
    a. a distinctive key bed in both areas.
    b. the same fossil content.
    c. the same rock type.
    d. any of the above

13. A good characteristic for a guide fossil is
    a. long life span as a species.
    b. a narrow geographic distribution.
    c. difficult identification.
    d. none of the above

14. An alpha particle is composed of
    a. two protons and two neutrons.
    b. a proton and an electron.
    c. a neutron and an electron.
    d. an electron.

15. The parent - daughter pair of U-235 and Pb-207 has a half-life of 704 million years. If a rock had 1 million atoms of U-235 when it formed and now has 250,000 atoms, how old is it?
    a. 704 million years
    b. 1.408 billion years
    c. 1 billion years
    d. 4.5 billion years

16. U-238 (Uranium) has an atomic # of 92 and atomic mass of 238. If it gives off 1 alpha particle and 2 beta particles it will become
    a. U-234 (Uranium: # of 92, mass of 234).
    b. Pb-206 (Lead: # of 82, mass of 206).
    c. Th-230 (Thorium: # of 90, mass of 230).
    d. Ra-226 (Radium: # of 88, mass of 226).

17. Of the list below, the rock type most easily dated radiometrically is
    a. granite.
    b. shale.
    c. limestone.
    d. All of the above equally easy.

18. Radioactive dates will be inaccurate if the sample has experienced
    a. leakage of the daughter isotope.
    b. heating during metamorphism.
    c. addition of parent atoms after its formation.
    d. any of the above

19. Radioactive carbon dates are useful for the past
    a. 5000 years.
    b. 70,000 years.
    c. 250,000,000 years.
    d. 4.5 billion years.

20. Under ideal circumstances, cross-dating techniques can extend tree ring dating
    a. only back to 600 years or so.
    b. back to about 14000 years.
    c. back to 220,000 years.
    d. back nearly seven million years.

21. The geologic time scale was developed using
    a. relative dating methods.
    b. absolute dating methods.
    c. a combination of a and b
    d. neither a nor b

22. A half-life is the time it takes for half
    a. of the daughters to become parents.
    b. of the parent isotopes to become trapped in a crystal.
    c. of the parent isotope to alter to daughter.
    d. of the daughter isotope to escape.

23. The great length of geologic time means
    a. even slow processes can cause great changes on the Earth.
    b. rare but powerful events like comet impacts are important.
    c. the Earth's geography has changed many times.
    d. all of these answers

24. Geologists demonstrate time-equivalency of rock units in different areas using
    a. correlation.
    b. Coriolis analysis.
    c. Corona procedures.
    d. theoretical conundrums.

25. When atoms decay
    a. their number of electrons change.
    b. their atomic number changes.
    c. the rate is partly dependent on pressure and temperature.
    d. they become ions.

TRUE OR FALSE

___1. Relative time tells us how long ago an event took place.

___2. The geologic time scale is based on rock sequences placed in chronological order.

___3. James Hutton's principle of uniformitarianism required that the earth must be older than anyone had previously thought.

___4. The principle of superposition allows geologists to obtain an absolute date for a given rock layer.

___5. If a fault crosscuts a rock layer, then the fault is younger than the rock layer.

___6. William Smith demonstrated that fossils succeed one another in time in a predictable order.

___7. A disconformity often requires the use of fossils for its recognition.

___8. A nonconformity exists when sedimentary beds underlie massive crystalline rocks.

___9. An angular unconformity exists when the beds above and below the unconformity have an angular relationship.

___10. The process of establishing the time equivalency of rock units in different areas is called correlation.

___11. Correlation on the basis of rock type is only reliable in a fairly restricted area.

___12. Guide fossils have a restricted geographic range and short time of existence.

___13. If one does not have guide fossils, one cannot use fossils for correlation.

___14. Although correlating rocks exposed at the surface is fairly easy, we have not yet devised methods for correlating rocks in the subsurface.

___15. After three half-lives a rock will have 1/3 of the atoms of radioactive parent product that it originally had.

___16. Only under unusual circumstances can the radiometric age for sedimentary rocks be determined.

___17. Metamorphism can reset a rock's radioactive "clock".

___18. A given rock type can only have one radioactive isotope.

___19. Even under ideal circumstances, error for radiometric dating is almost never better that 15% of its age.

____20. One advantage of fission track dating is that it is one of the few ways to absolutely date the Pleistocene rocks.

____21. Carbon-14 dates have a short range of application because the half-life of carbon-14 is very short.

____22. Carbon-14 enters the bodies of living organisms in a constant ratio to other isotopes of carbon throughout that organism's life.

____23. Relative dating uses logical reasoning to analyze the order of geologic events.

____24. Fission track dating is most useful for rocks over 1.5 billion years old.

____25. Whenever possible rocks are dated using more than one set of radioactive isotope pairs.

DRAWINGS AND FIGURES

1. The Grand Canyon, with its spectacularly eroded sedimentary rocks, is illustrated in the photograph below. Answer the following questions while studying the photo.
   a. What principle suggests that the layers have been undisturbed since they were deposited?
   b. What principle suggests that A, B and C are part of the same layer even though deep canyons now separate these locations?

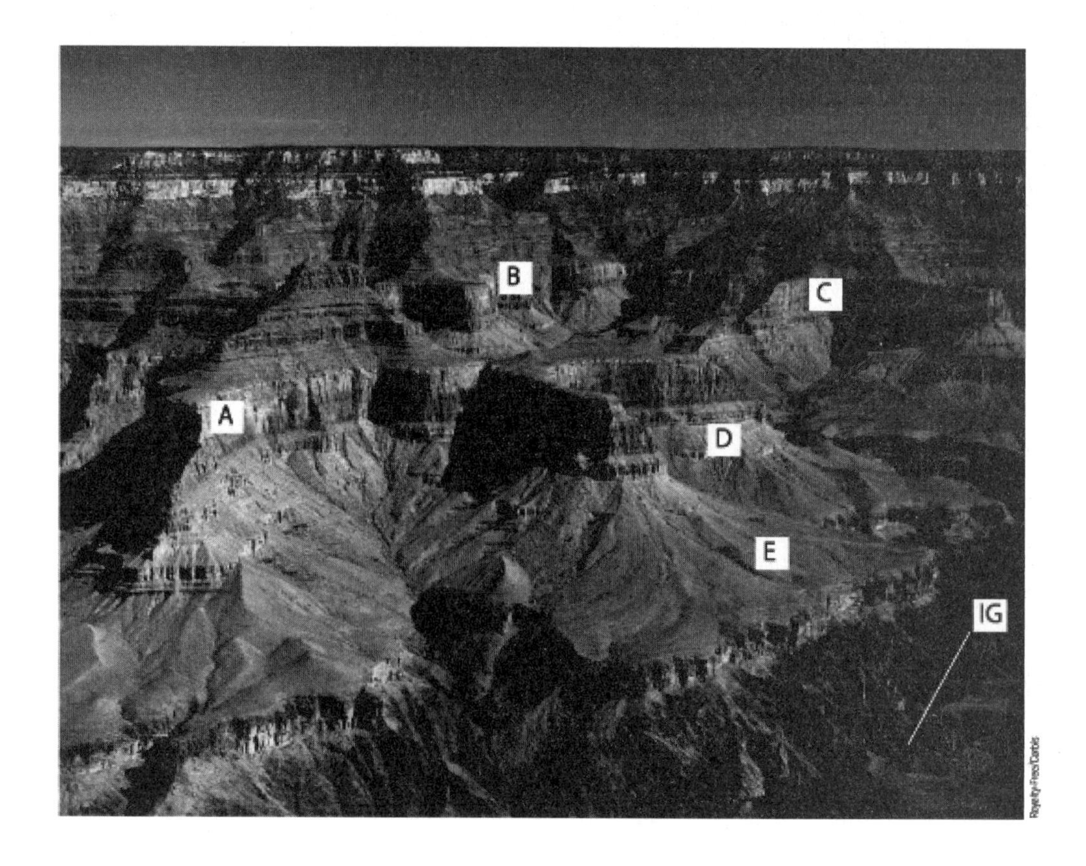

2. Sketch cross-sectional views of each of the three types of unconformities. Be sure to label rock type.

3. Using the basic principles of relative dating, list the rocks, unconformities and faults labeled in the block diagram below in order from oldest to youngest.

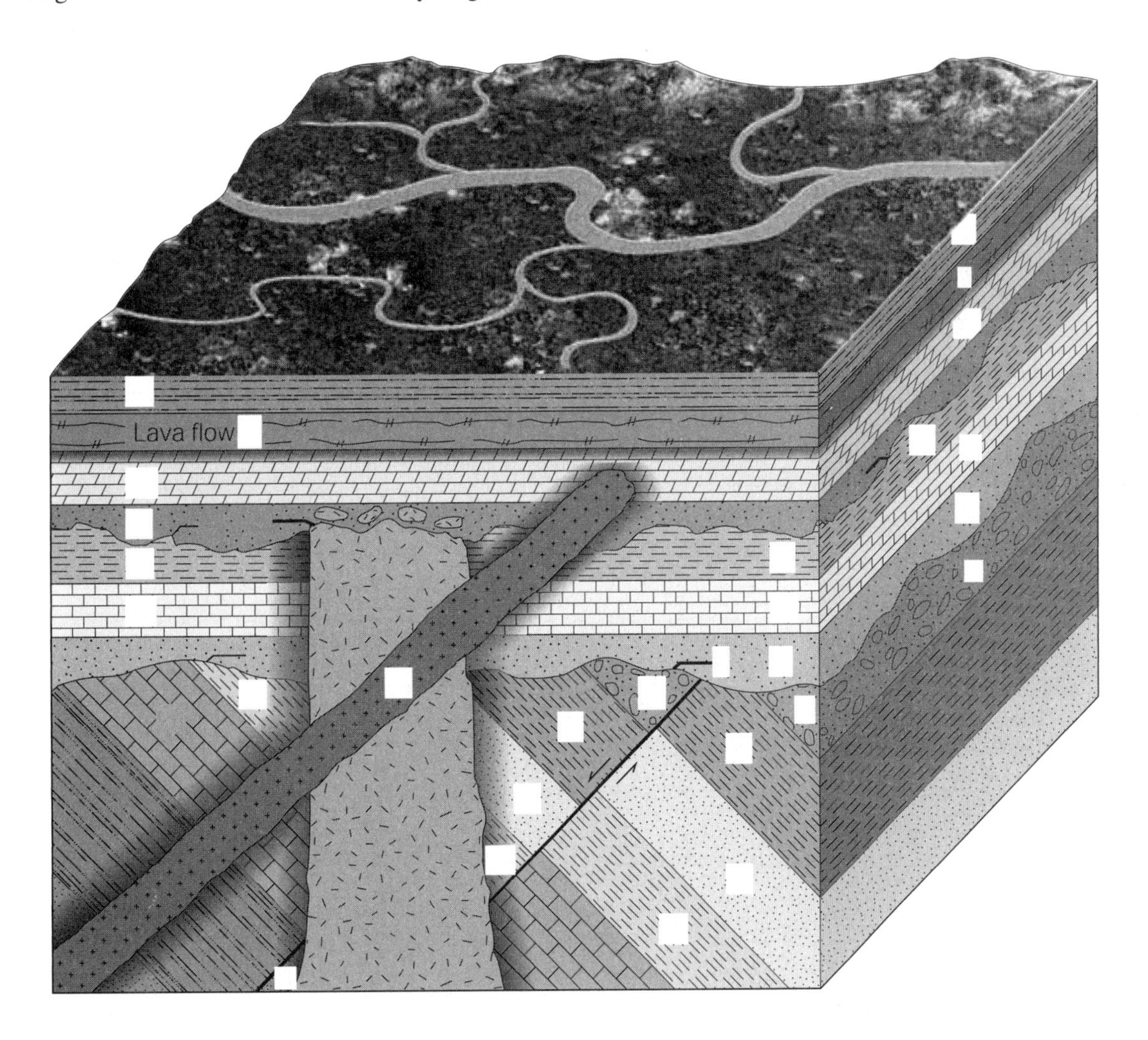

4. The diagram below illustrates the range of time over which five species existed. If you find a rock that contains all five as fossils, what is the oldest and what is the youngest possible age the rock can be determined by using concurrent range zones.

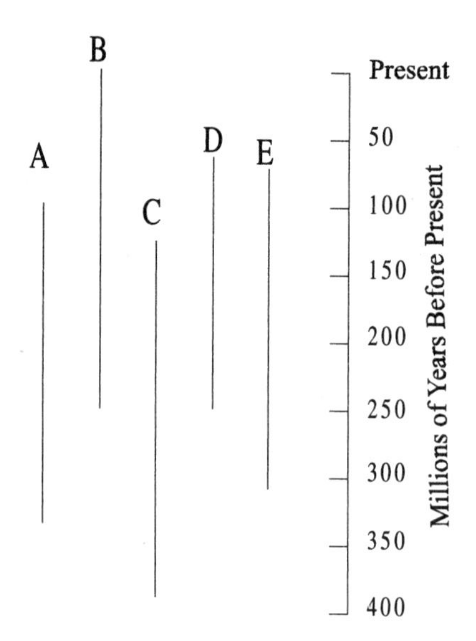

Chapter 10

Earthquakes

CHAPTER OBJECTIVES

By the end of this chapter you should be able to:

1. Explain the elastic rebound theory for earthquakes.
2. Explain how a simplified modern seismograph works.
3. Differentiate between shallow, intermediate and deep focus earthquakes.
4. Describe where earthquakes occur and why they occur where they do.
5. Compare and contrast the various types of seismic waves including such aspects as relative velocities and direction of particle movement.
6. Determine the distance from a seismic station to an earthquake epicenter.
7. Determine the precise location of an earthquake epicenter.
8. Explain how earthquake intensity is determined.
9. Using the appropriate scale, determine the magnitude of an earthquake.
10. Produce a comprehensive list of damaging effects of earthquakes.
11. Define what is meant by "earthquake precursor" and give several examples.
12. Discuss the prospects for earthquake control.

USEFUL ANALOGIES

A Slinky® is a useful tool for demonstrating P and S waves. Stretch it out between two people. Let one person be the wave source. When the wave source gives a snap of the wrist, a wave will travel down the slinky, This wave will "vibrate" perpendicular to its direction of travel. After that, have the wave source give a sharp push parallel to the Slinky®. This push must be very sharp and quick. This time a wave will travel down the Slinky® but it will vibrate parallel in the same direction of travel being manifested as a compression and extension of the spring as it travels. Note which of the two migrates the fastest.

KEY TERMS

After reading this chapter , you should be familiar with the following terms.

earthquake
aftershock
elastic rebound theory
seismology
seismograph
seismic waves

seismogram
focus
hypocenter
epicenter
shallow focus
intermediate focus

deep focus
Benioff zone
Circum-Pacific belt
Mediterranean-Asiatic belt
body waves
surface waves
P-waves (primary waves)
S-waves (secondary wave)
shear waves
elasticity
surface waves
R-waves (Rayleigh waves)
L-waves (Love waves)
time-distance graph
intensity
magnitude
Modified Mercalli Intensity Scale
Richter Magnitude Scale
Moment Magnitude Scale
liquefaction
tsunami
seismic risk maps
precursors
seismic gaps
tiltmeter
dilatancy

CHAPTER CONCEPT QUESTIONS

1. Explain the elastic rebound theory as a cause of earthquakes.

2. What is the difference between the epicenter of an earthquake and the focus (hypocenter)?

3. Explain how earthquakes are distributed relative to plate boundaries. Where do deep and intermediate earthquakes occur and why?

4. What is the difference between body and surface waves?

5. List the four types of seismic waves in order of arrival at a seismic station (i.e. decreasing velocity). For each type describe the "vibration direction" or direction of material movement relative to the direction of wave travel.

6. How is the distance from a seismic station to an earthquake epicenter determined?

7. How do seismologists precisely locate an earthquake epicenter?

8. How is earthquake intensity determined? What are these values used for?

9. How is earthquake magnitude determined? How does it relate to energy released during an earthquake?

10. How does the ground material upon which a structure is built influence the extent to which it might be affected by ground shaking during and earthquake?

11. What sort of precautions should developers in earthquake-prone areas take to mitigate the potential damages from earthquakes?

12. Besides shaking of the ground, what are some potential destructive effects of earthquakes?

13. How does a tsunami form? What signs along a coast might suggest that one is approaching.

14. Describe four earthquake precursors.

15. What are seismic gaps and how are they used in earthquake prediction?

16. Describe a possible technique for controlling earthquakes that was suggested by experiments in old Colorado oil fields?

COMPLETION QUESTIONS

1. An earthquake is a shaking of the earth, which occurs when ________ stored in rocks is released, causing displacement of those rocks along a fault.

2. The favored explanation for the origin of earthquakes is called the ________ ________ theory.

3. The study of earthquakes is called ________.

4. The instrument used to record seismic waves is a ________ and the record made by this instrument is called a ________.

5. The point of origin of an earthquake in the subsurface is termed the __________, and the point on the surface directly above is called the ________.

6. Earthquakes classified as "shallow" have foci located less than _____ km below the surface. Those classified as "deep" have foci located over _____ km below the surface.

7. Focal depths increase along a dipping zone beneath convergent plate margins called ________ zones.

8. 80% of all earthquakes occur in the ________ belt.

9. The ________ belt accounts for about 15% of all earthquakes.

10. The number of earthquakes recorded each year by seismographs is ________.

11. Seismic waves that travel through the Earth are called ________ waves, and those that travel along the ground surface are called ________ waves.

12. Two types of body waves exist, ________ waves ________ waves.

13. The fastest moving waves, ________, cause earth material to move back and forth ________ to the direction of wave motion.

14. ________ are the second type of waves to arrive at a seismic station. They cause earth materials to move up and down ________ to the direction of travel.

15. The two most important types of surface waves are ________ and ________ waves.

16. By measuring the difference between the arrival times of ________ and ________ waves, the distance to, and the location of an earthquake can be determined.

17. Earthquake ________ is based on building damage and personal observations and is measured according to the ________ scale.

18. Earthquake ________ is based on energy released during an earthquake and is measured according to the ________ scale.

19. A seismogram showing the record of an earthquake of magnitude 5 has a peak with ____ times the amplitude of one recorded during an earthquake of magnitude 3.

20. In an earthquake, ground shaking is the most severe in areas of loose, unconsolidated and waterlogged material when the ground starts behaving as a fluid. This process is called _______.

21. As a result of an underwater earthquake, coastal regions could be devastated by a _________.

22. _________ maps record the distribution of previous earthquakes in the hope that they will indicate the likelihood of future severe quakes.

23. In an effort to predict earthquakes, geologists are trying to identify changes in the Earth which precede earthquakes, called _________ , but nothing consistent or reliable has yet been discovered.

24. The majority of earthquakes are ______ - focus quakes, which are almost always the ______ destructive.

25. The farther seismic waves travel from their source, the ________ the shaking felt and the ______ the gap between P and S waves allowing scientists to locate the epicenter.

MULTIPLE CHOICE

1. When a material returns to its original shape after the release of stress it is said to be
   a. ductile.
   b. brittle.
   c. elastic.
   d. none of these

2. The instrument that detects and records seismic waves is called a
   a. seismograph.
   b. seismologist.
   c. seismogram.
   d. none of these

3. Approximately 90% of all earthquakes are ______ focus.
   a. shallow
   b. intermediate
   c. deep
   d. out of

4. The point on the surface, directly above the subsurface location where an earthquake originates is called the
   a. focus.
   b. hypocenter.
   c. epicenter.
   d. all of the above

5. The most destructive earthquakes have
   a. shallow focus.
   b. intermediate focus.
   c. deep focus.
   d. All of the above are equally destructive.

6. _____ focus earthquakes occur at convergent margins.
   a. Shallow
   b. Intermediate
   c. Deep
   d. all of the above

7. Most of the world's earthquakes occur
   a. in the Circum-Pacific belt.
   b. in the Mediterranean-Asiatic belt.
   c. in the interior of plates.
   d. They are equally distributed around the earth.

8. Earthquakes occur where plates
   a. slide past each other.
   b. converge.
   c. diverge.
   d. all of the above

9. S-waves will travel through
   a. liquids.
   b. solids.
   c. air.
   d. all of the above

10. The ranking of seismic waves from fastest to slowest is
    a. L,S,P.
    b. P,S,L.
    c. P,L,S.
    d S,P,L.

11. Factors that control seismic wave velocity include
    a. density and dilatancy.
    b. elasticity and resistivity.
    c. density and elasticity.
    d. pressure and dilatancy.

12. The intensity scale is
    a. based on building damage and personal observation.
    b. based on the amplitude of seismic waves.
    c. based on the magnitude of seismic waves.
    d. another name for the magnitude scale.

13. The intensity of an earthquake is controlled
    a. by the size and duration of the earthquake.
    b. the depth of the focus, and distance from the epicenter.
    c. by the local geology.
    d. all of the above

14. A building will sustain the least amount of damage if it is built on
    a. artificial fill.
    b. water saturated sediments.
    c. unconsolidated sediments.
    d. bedrock.

15. An earthquake's magnitude takes into account
    a. the amount of building damage.
    b. the amount of energy released by the earthquake.
    c. the number of people displaced.
    d. all of the above

16. The Richer scale uses the _________ peak on a seismogram and so represents an instant in time. This means that the total energy of the earthquake is __________.
    a. highest / overestimated
    b. highest / underestimated
    c. lowest / overestimated
    d. highest / underestimated

17. An earthquake of magnitude 4 releases ______ times the energy as one of magnitude 2.
    a. 2
    b. 60
    c. 100
    d. 900

18. Which scale provides insight into the size of earthquakes that occurred before humans were around to witness and record them.
    a. the Modified Mercalli Scale
    b. the Richter Magnitude Scale
    c. the Moment Magnitude Scale
    d. all of the above

19. Tsunamis are large destructive waves that are due to
    a. earthquakes.
    b. submarine landslides.
    c. volcanic eruptions.
    d. all of the above

20. By far most of the damage caused by the great 1906 San Francisco earthquake resulted from
    a. ground shaking.
    b. ground rupturing.
    c. fire.
    d. tsunami.

21. Events that geologists monitor to try and predict earthquakes include
    a. seismic gaps.
    b. changes in elevation and tilting of the ground.
    c. fluctuations of water levels in water wells, variations in the Earth's magnetic and electric resistance.
    d. all of the above

22. The Earth experiences an average of ________ earthquakes per year.
    a. 10
    b. 150,000
    c. 1,050,000
    d. 90,000

23. Seismic waves
    a. move outward in all directions from their source.
    b. move in a north-south direction only.
    c. travel at different velocities depending on the weather.
    d. travel faster in continental crust.

24. Which of the following does NOT control the intensity of an earthquake?
    a. local geology
    b. length of shaking
    c. weather, especially amount of heat
    d. distance form epicenter

25. The factors that control the number of people killed in an earthquake include
    a. population density.
    b. building design and contruction.
    c. earthquake focal depth.
    d. all of these answers

TRUE OR FALSE

____1. A seismogram is the instrument used to record seismic waves.

____2. The epicenter of an earthquake is always located directly over the focus.

____3. Earthquakes along a transform plate boundary generally have a deep focus.

____4. The San Andreas Fault is part of the circum-Pacific belt where most of the world's earthquakes occur.

____5. Most earthquakes have a deep focus.

____6. Earthquakes occurring in the interior of a plate, far from boundaries are never large or dangerous.

____7. Seismographs may still detect earthquakes too weak to be felt by people.

____8. P-waves will not travel through liquids.

____9. P-waves are the most damaging type of earthquake waves to building foundations.

____10. In S-waves, earth materials move up and down, perpendicular to the direction of movement of the wave.

____11. The distance from a seismograph to the epicenter of an earthquake is determined by using the difference in arrival times between P and S waves.

____12. The precise location of an earthquake can be determined with as few as two seismograph stations.

____13. A well made, accurate seismograph is fundamental to determining earthquake intensity.

____14. Insurance companies are more concerned with magnitude than intensity.

____15. An isoseismal line is a line of equal magnitude.

____16. A one-unit increase in Richter magnitude represents a 10-fold increase in earthquake size.

____17. Analysis of earthquake damage has repeatedly shown that ground shaking is most severe in areas underlain by bedrock.

____18. On a worldwide basis, there are more small earthquakes each year than large ones.

____19. An earthquake of magnitude 6 releases twice as much energy as a magnitude 3 earthquake.

____20. An earthquake of intensity X will register 10 times the amplitude on a seismogram as one of intensity IX.

____21. Statistically, more people die in an earthquake that occurs at night when people are asleep.

___22. Coastal areas thousands of miles from an earthquake can be affected by a tsunami.

___23. Our ability to predict earthquakes is now quite good with long and short range prediction accuracy close to 90%.

___24. Deliberately causing small earthquakes by pumping water into fault zones in order to prevent larger events has gained considerable popularity and is widely practiced in earthquake-prone areas.

___25. Larger earthquakes shake longer because a longer section of the fault rupture.

DRAWINGS AND FIGURES

1. On the figure below, highlight the area in which you would expect find foci (a) shallow, (b) intermediate, and (c) deep focus earthquakes.

2. Using the graph to the right, determine the distance between your seismograph station and the epicenter of an earthquake that registers 5 minutes between the arrival of P and S waves on your seismograph.

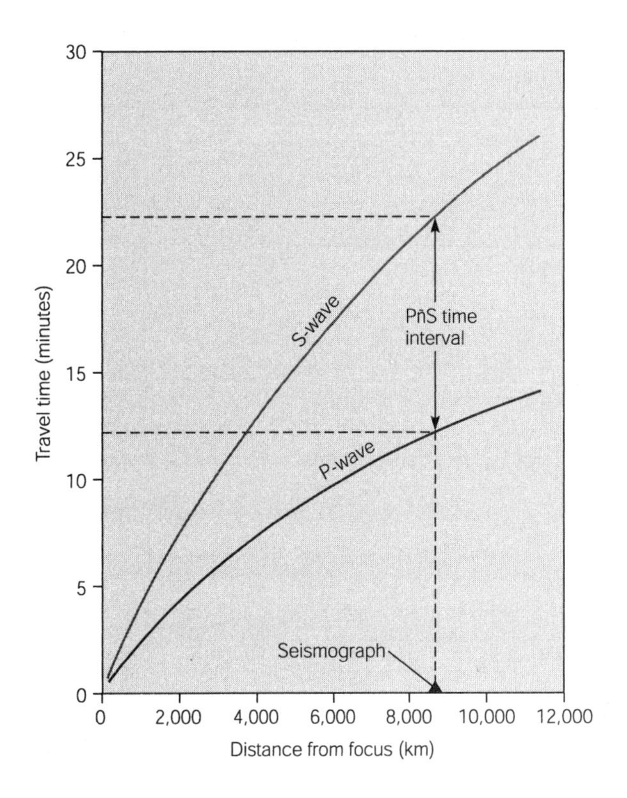

3. You are 20 km away from an earthquake which shows an amplitude of 100 mm on your seismogram. Using the graph below, determine its magnitude.

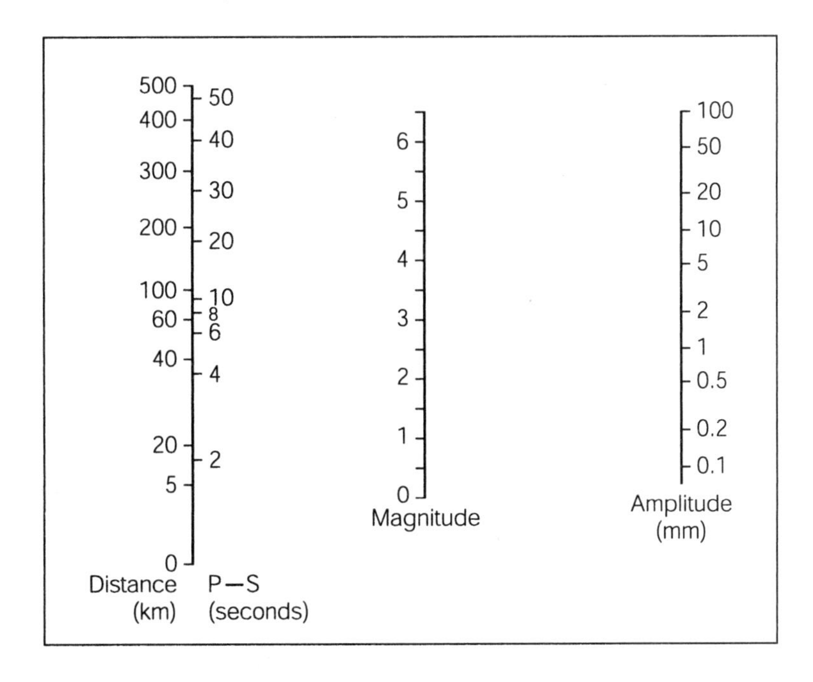

Chapter 11

Earth's Interior

CHAPTER OBJECTIVES

By the end of this chapter you should be able to:

1. Describe changes in velocity and direction of seismic waves as they travel through the Earth and the reasons for these changes.
2. Summarize the information that seismic tomography has contributed to our model of the mantle.
3. Explain why scientists think that the Earth has both a solid and liquid core.
4. Characterize each layer of the Earth in terms of composition, density, percentage of total Earth volume, and percentage of total Earth mass.
5. Differentiate between the distinct zones recognized in the mantle and characterize each.
6. Account for the variations in thickness of the crust.
7. Recall the following values: the geothermal gradient in (a) the crust and (b) in the mantle; the estimated temperature at (c) the base of the crust, (d) the core-mantle boundary, and (e) the center of the core.
8. List the areas of higher and lower than average heat flow.
9. Explain why the gravitational attraction between a suspended mass and the earth varies from place to place on the Earth.
10. Explain what isostasy is and why it occurs.
11. Differentiate between inclination and declination.
12. Explain what magnetic anomalies are and why they occur.
13. Explain what magnetic reversals are and what record they leave in the rocks.

KEY TERMS

After reading this chapter you should be familiar with the following terms.

wave rays
refraction
reflection
discontinuity
seismic tomography
P-wave shadow zone
S-wave shadow zone
core
mantle
Mohorovicic discontinuity (Moho)
low velocity zone
asthenosphere
lithosphere
transition zone
peridotite
kimberlite pipes
continental crust
oceanic crust
geothermal gradient
heat flow
law of universal gravitation
gravity

weight
gravimeter
positive gravity anomaly
negative gravity anomaly
principle of isostasy
isostatic rebound
magnetic field
dipolar
Curie point
magnetite
self-exciting dynamo
magnetic inclination
magnetic declination
magnetic anomalies
magnetometer
positive magnetic anomaly
negative magnetic anomaly
paleomagnetism
magnetic reversals
normal polarity
reverse polarity

CHAPTER CONCEPT QUESTIONS

1. Why do seismic waves refract and reflect?

2. What is seismic tomography? What has it contributed to our models of the mantle and core?

3. What is the evidence that the outer core is liquid? What is the evidence that there is a solid inner core?

4. Any model on the origin of the core must account for what two phenomena related to the core? How are these phenomena accounted for in currently favored models?

5. What is the Moho and how was it discovered?

6. What causes the sudden decrease in velocity in the *low velocity zone*?

7. What is believed to be the cause of the discontinuities of the transition zone?

8. Compare and contrast oceanic and continental crust in terms of composition, thickness and density.

9. Why do geophysicists think that the geothermal gradient in the mantle is much less than in the crust?

10. What factors determine the force of gravity? Why is the gravitational attraction between the Earth and moon so much more than between Earth and a small space craft an equal distance away? Why is the gravitational pull of the Earth on the spacecraft higher when it is in our atmosphere than when it is beyond our atmosphere?

11. How does the gravitational attraction of the Earth on a gravimeter over an iron ore body compare to the attraction over a salt dome and why?

12. Why does a mountain range rise vertically as it is eroded?

13. What is the difference between the attractive force of gravity and that of magnetism?

14. Why don't geologists think that the Earth's magnetic field is due to a large body of magnetic material, such as magnetite, below the surface? What alternative origin for the magnetic field has been favored?

15. What is the difference between magnetic inclination and declination?

16. Why do iron-rich rocks produce a positive gravity anomaly?

17. What is the evidence that the Earth has reversed magnetic polarity from time to time?

COMPLETE THE FOLLOWING TABLE:

LAYER	COMPOSITION	DENSITY	% TOTAL VOLUME	% TOTAL MASS
Inner Core				
Outer Core				
Mantle				
Cont. Crust				
Oceanic Crust				

COMPLETION QUESTIONS

1. The average density of Earth has been calculated to be close to______, but the average surface rock has a density between _______ and ______.

2. Knowledge of the internal structure of the Earth is derived primarily from the behavior and travel times of ________.

3. As seismic waves move through the Earth, they refract when entering material of different ________ and ________.

4. Refraction involves a change in a wave's ________ and ________.

5. When a seismic wave reaches a boundary separating materials of significantly different ________ and ________, some of the energy is ________ back to the surface.

6. __________ has yielded a clearer three dimensional picture of Earth's interior revealing fast and slow areas, perhaps related to convection.

7. __________ has shown that the core/mantle boundary is not smooth, but has rises and depressions.

8. The nature of the core was postulated from the observations of P- and S-wave ________ zones.

9. P-wave shadow zone extends between ____ and ____ degrees in either direction measured from an earthquake's ________.

10. S-waves will not reach a seismograph station located more than ______ degrees from the focus.

11. The outer core is molten, probably largely due the presence of __________, which depresses the melting temperature.

12. Between the core and the crust is the iron-magnesium rich ________ which comprises more than 80 percent of the Earth's volume.

13. Marked changes in velocity of seismic 20 to 90 km below continental crust and 5 to 10 km below oceanic crust which indicates the boundary between the crust and the mantle called the ________.

14. A discontinuity within the mantle, at a depth of about 410 km., has been attributed to changes in ________ structure.

15. Peridotite inclusions are found in ___________ which are volcanic bodies that originated in the mantle.

16. The outermost layer, called the crust, is divided into the thinner ________ crust and the thicker ________ crust.

17. In very broad terms, the ocean crust is ________ and the continental crust is ________ in composition.

18. The geothermal gradient averages ________ per kilometer through the crust.

19. Gravitational force between two bodies ________ with increasing mass and ________ with increasing distance.

20. A sensitive instrument called a ________ detects variations in gravitational attraction of the earth at different locations.

21. If the gravitational attraction is higher or lower than expected for a given elevation, then a ________ gravity anomaly is said to exist.

22. The principle of ________ explains how Earth's crust floats on the ________.

23. As a result of isostatic equilibrium, the addition of weight on the Earth's crust causes it to ________, while the removal of weight from the Earth's surface cause the crust to ________.

24. When rocks are heated to their ________ point they lose their magnetism.

25. The Earth's magnetic field appears to be due to a self exciting dynamo produced by convection in the ________.

26. With time, the strength of the magnetic field weakens, and the ________ of the field can reverse.

27. Deviation of the magnetic field from the horizontal is called magnetic ________ and the angle between geographic north and magnetic north is called the ________.

28. Variations in the strength of the magnetic field are called magnetic ________.

29. ________ is the study of remnant magnetism in ancient rocks. It records the direction and strength of the magnetic field at the time of rock formation.

30. Rocks that have a record of magnetism the same as today's are described as having ________ polarity, whereas rocks with the opposite magnetism have ________ polarity.

MULTIPLE CHOICE

1. Generally, the velocity of seismic waves traveling through the mantle _______ with depth.
   a. increases
   b. decreases
   c. fluctuates randomly
   d. does not change

2. As P-waves travel from the outer core to the mantle, they
   a. increase in velocity.
   b. decrease in velocity.
   c. disappear entirely.
   d. are all reflected back to the surface.

3. Geologists know that the outer core is liquid because
   a. P-waves will not travel through this region.
   b. L-waves will not travel through this region.
   c. S-waves will not travel through this region.
   d. all of the above

4. Much of the core is thought to be composed of a high-pressure allow of __________ and __________.
   a. iron and nickel
   b. iron and magnesium
   c. iron and sulfur
   d. silicon and aluminum

5. The Moho is the boundary between
   a. the mantle and the core.
   b. the crust and the core.
   c. the upper mantle and the lower mantle.
   d. none of these

6. The low velocity zone roughly corresponds to the
   a. lithosphere.
   b. asthenosphere.
   c. outer core.
   d. inner core.

7. The mantle is believed to be largely composed of the rock _________.
   a. gabbro
   b. granite
   c. basalt
   d. peridotite

8. In broad, general terms, the
   a. oceanic crust is granitic, and the continental crust is basaltic.
   b. oceanic crust is ultramafic, and the continents are granitic.
   c. oceanic crust is basaltic, and the continents are granitic.
   d. none of the above

9. The continental crust is _________ the oceanic crust.
   a. thicker and less dense than
   b. thinner and less dense than
   c. thicker and denser than
   d. equal in thickness and density to

10. Most of the heat lost by the Earth (about 70%) is lost through the
    a. high mountains of the continents.
    b. the flat areas of continents.
    c. the sea floor.
    d. the three deepest ocean trenches.

11. Force of gravity between two bodies
    a. increases as the masses decrease.
    b. increases as the masses move farther apart.
    c. increase as centrifugal force increases.
    d. none of the above

12. The force of gravity varies with
    a. latitude and longitude.
    b. elevation and latitude.
    c. longitude and elevation.
    d. none of these

13. Over a buried mass of iron ore, gravity measurements would
    a. be lower than over average crust.
    b. be equal to that of average crust.
    c. be higher than average crust.
    d. could not be made due to magnetic interference.

14. Gravimeter readings across mountains show
    a. a high gravity anomaly.
    b. a low gravity anomaly.
    c. either **a** or **b** depending on latitude.
    d. no gravity anomaly.

15. Which of the following would be expected to yield a negative gravity anomaly?
    a. an ore body
    b. a mountain range
    c. a salt dome
    d. all of the above

16. An example of isostatic rebound is
    a. the rise of continental crust as glaciers melt.
    b. the rise of mountains even as they erode.
    c. the maintenance of 10% of an iceberg above the surface of the water no matter how much melts.
    d. all of the above

17. The Earth's magnetic field is most likely due to
    a. changes in the rotation of the Earth.
    b. a highly magnetic core.
    c. electrical currents in the outer liquid core.
    d. electrical currents in the mantle.

18. Deviations from the normal magnetic field strength are called
    a. magnetic inclinations.
    b. magnetic declinations.
    c. magnetic anomalies.
    d. none of these

19. Magnetic reversals occur when Earth's
    a. polarity reverses.
    b. magnetic declination increases.
    c. magnetic inclination decreases.
    d. none of these

20. Over the last century the Earth's magnetic field has
    a. decreased in strength.
    b. increased in strength.
    c. stayed the same.
    d. reversed polarity.

21. Scientists study the Earth's interior by
    a. studying the behavior of seismic waves.
    b. studying meteorites.
    c. laboratory experiments.
    d. all of these answers

22. The core-mantle boundary
    a. is called the Moho.
    b. is the hottest part of the Earth.
    c. is NOT a smooth sphere.
    d. marks the boundary between iron and nickel layers.

23. Which layer in the Earth has the highest density and accounts for one third of Earth's mass?
    a. core
    b. lithosphere
    c. mantle
    d. ocean crust

24. Velocities of body waves are controlled by
    a. density and elasticity of rock.
    b. weather.
    c. strength of the earthquake.
    d. type of plate margin.

25. Gravimeters
    a. can be used to find ore deposits.
    b. can detect rock density differences in the subsurface.
    c. have been used to find salt domes.
    d. all of these answers

TRUE OR FALSE

____1. The Earth's average density is considerably greater than that of the rocks in the crust.

____2. Wave velocity increases with increasing elasticity.

____3. Seismic tomography has provided evidence of convection cells in the mantle.

____4. Seismic velocities are slower in hot parts of the mantle than in cooler parts.

____5. The S-wave shadow zone is due to the refraction of waves as they travel through the inner core.

____6. The core is composed of similar minerals to those found in the crust.

____7. The Moho is the boundary between the outer and inner core.

____8. The transition zone roughly corresponds to the asthenosphere.

____9. Once below the asthenosphere, the mantle is homogeneous.

____10. The mantle is composed mostly of gabbro.

____11. The continental crust is being stretched and thinned in areas like the East African Rift Valley and the Basin and Range Province of the U.S.

___12. Oceanic crust is thickest at the spreading ridges.

___13. Most of Earth's internal heat is generated by radioactive decay in the core.

___14. The temperature in the Earth's core is thought to be close to the temperature of the surface of the sun.

___15. Heat flow is highest at the spreading ridges.

___16. An object weighs slightly less at the equator than it would at the poles.

___17. The density of the material beneath the ground can affect the force of gravity in a given spot.

___18. A huge positive gravity anomaly exists under high mountain ranges such as the Himalayas.

___19. Since the last ice age much of the Northern Hemisphere has been subsiding.

___20. The Earth is an example of a dipolar magnetic field.

___21. Heating magnetic minerals to their Curie Point strengthens or enhances their magnetism.

___22. The 11.5 degrees separating the magnetic north pole from the geographic North Pole is called the magnetic inclination.

___23. Iron-bearing geologic materials that add to the measured magnetic field produce positive magnetic anomalies.

___24. During a period of magnetic reversal, the north arrow on a compass needle would point south.

___25. Basaltic lava flows are very useful rocks in preserving paleomagnetism.

DRAWINGS AND FIGURES

1. In the figure to the right label the layers of the Earth. In the inset in the upper right label the lithosphere, asthenosphere, upper mantle and the two types of crust.

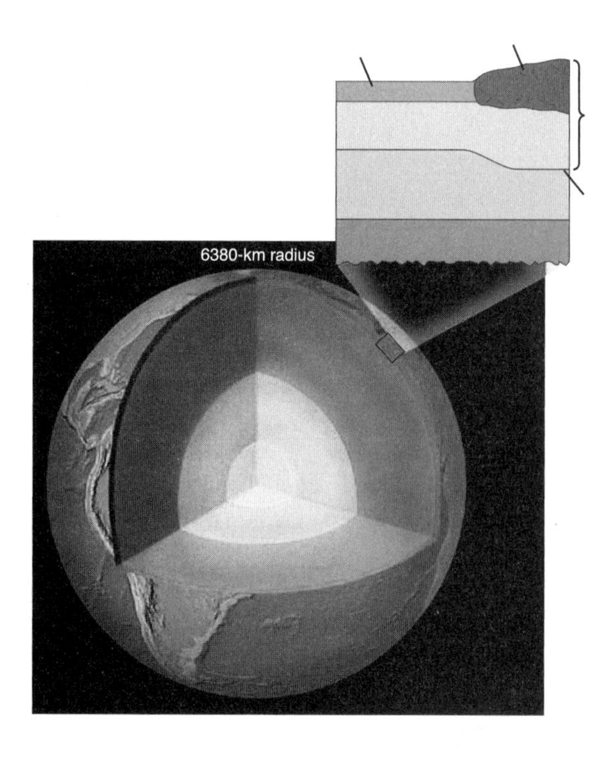

2. On the graph of P and S wave velocity vs. depth, indicate the location of (a) the crust-mantle boundary, (b) boundary between the outer core and mantle, and (c) the low velocity zone.

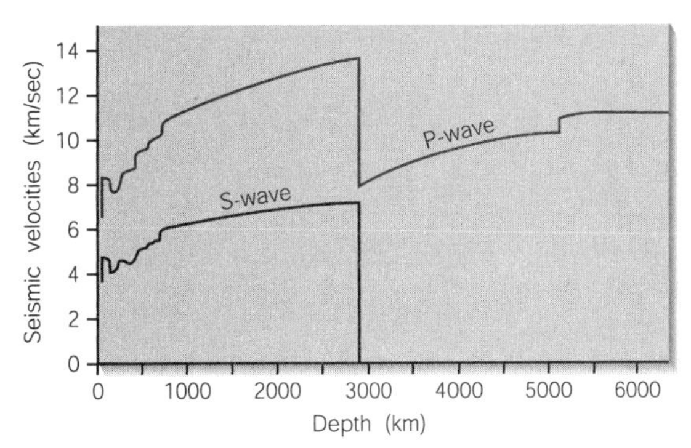

3. The adjacent figure shows the various paths of seismic waves through the earth from an earthquake focus. Label the P and S wave shadow zones.

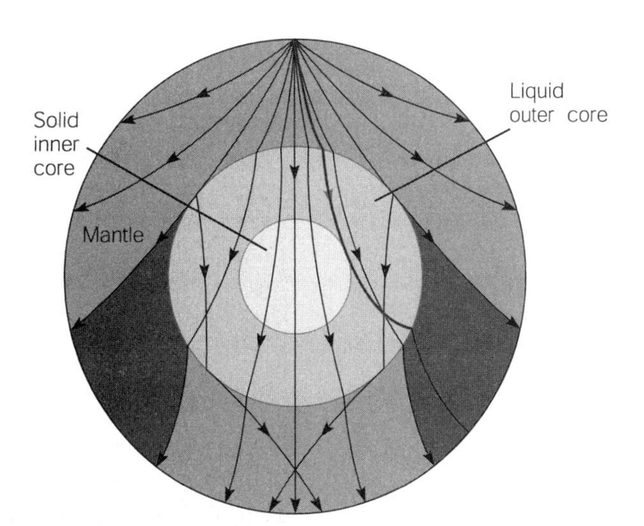

4. In the three diagrams below, A and B are stations at which magnetic and gravity readings are made. At each station indicate positive and negative gravity.

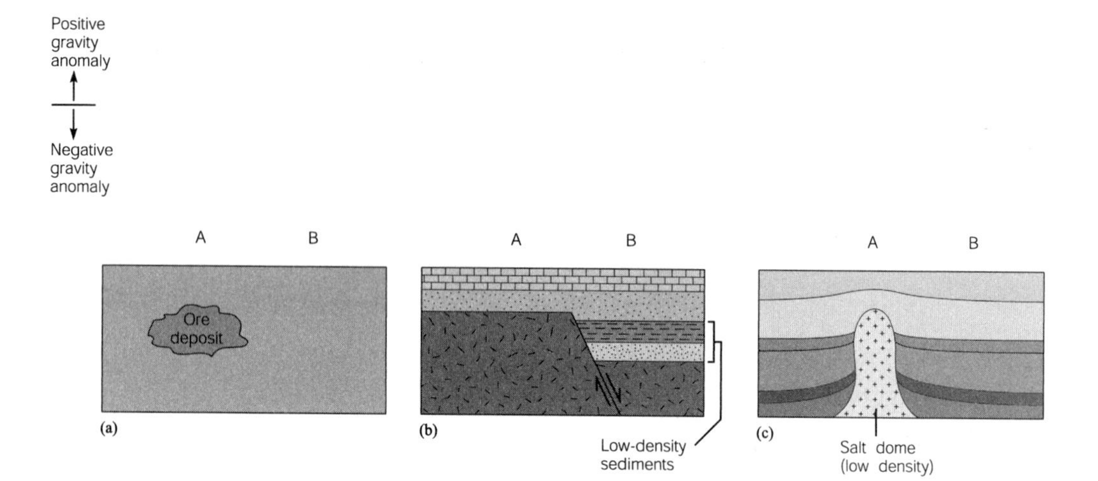

Chapter 12

The Seafloor

CHAPTER OBJECTIVES

By the end of this chapter you should be able to:

1. Compile a list of tools used in modern oceanographic research.
2. Describe a vertical profile through the sea floor and tell what evidence has lead to the development of that model.
3. Describe the parts of a continental margin and classify continental margins by type.
4. Discuss the characteristics and effects of turbidity currents.
5. Describe the major features of the deep sea floor and give examples.
6. Characterize the types, and give the sources of, the various types of deep-sea sediment.
7. Recount the evolution of a reef.
8. List several economically important mineral resources that are found in the oceans.

KEY TERMS

After reading this chapter, you should be familiar with the following terms.

H.M.S. *Challenger*
echo sounder
seismic profiling
Deep-sea Drilling Project
Glomar Challenger
Alvin
ophiolite
continental margins
continental shelf
continental slope
shelf-slope break
continental rise
turbidity currents
graded bedding
submarine fans
submarine canyons
active continental margin
passive continental margin
abyssal plains
oceanic trenches
oceanic ridge
submarine hydrothermal vents
black smokers
seamounts
guyot
abyssal hills
aseismic ridges
microcontinents
manganese nodules
pelagic
pelagic clay
ooze
calcareous ooze
siliceous ooze
reefs
fringing reefs
barrier reefs
atoll
Exclusive Economic Zone
phosphorite

CHAPTER CONCEPT QUESTIONS

1. Briefly summarize the contributions to our understanding of the oceans of the following: The deep-sea Drilling Project, submersibles such as Alvin, echo sounding, and seismic profiling.

2. What are ophiolites? What do they contribute to our knowledge of the oceanic crust?

3. Compare and contrast the continental rise, slope and shelf in terms of their slopes, depths, and sedimentary processes.

4. How does a turbidity current work? What is the evidence that they exist?

5. Compare and contrast the passive and active of continental margins.

6. Even though the sea floor is rough when it is created at the ridge, the abyssal plains are flat. Explain why.

7. Summarize the characteristics of the oceanic trenches. Include information on typical depths, slopes, heat flow, gravity anomalies, and seismic activity.

8. Summarize the characteristics of the oceanic ridges.

9. Compare and contrast seamounts, guyots, abyssal hills, and aseismic ridges.

10. List the various types of sediment found in the deep sea and state the origin of each.

11. What resources does the ocean currently provide? What resources may be supplied in the future? What special problems does extraction of these resources pose?

COMPLETION QUESTIONS

1. During the 15 years of the Deep-sea Drilling Project, the ship named________ drilled over 1000 holes in the sea floor.

2. Some aspects of the subsurface geology of the sea floor are determined by ________ profiling.

An ________ is a sliver of sea floor emplaced on continental crust above subduction zones.

4. Continental margins consist of the continental ________, the continental ________, and continental ________.

5. In some places, the shelf and slope have been eroded, forming submarine ________.

6. Density currents called ________ currents have transported sediment from the continental shelf and slope to the deep sea floor. This sediment has largely accumulated in overlapping ________ at the continental ________.

7. Turbidites may be deposited beyond the continental rise to form the ________ plains.

8. ________ continental margins occur on the edge of a plate where oceanic crust is subducted in a deep oceanic ________.

9. Narrow continental ________ and the lack of a continental ________ characterize active continental margins.

10. The eastern coast of North America is a(n) ________ continental margin and the western coast is a(n) ________ continental margin.

11. ________ form some of the flattest, most featureless parts of the earth's surface.

12. The deepest parts of the oceans are found in the deep-sea ________.

13. The mid-oceanic ridges form a mountain range that extends at least ________ km around the globe.

14. Running along the crests of some oceanic ridges are ________ formed in response to tensional forces.

15. Hot, magma-heated water is discharged along the mid-ocean ridges through vents called ________

16. Adjacent segments of ocean ridges are offset by sea floor ________.

17. Here and there, rising over a kilometer above the flat abyssal plain are ________. Those with a flat top are called ________.

18. Ocean ridges that lack earthquakes are called ________ ridges.

19. Deep-sea sediments include large amounts of fine-grained material derived from the continents and islands called ________ clay.

20. ________ is the name given to sediments rich in the shells of microscopic plankton. These deposits can be ________ or siliceous in composition.

21. ________ are structures built by the skeletons of colonial organism.

22. In an island setting reefs evolve from ________ reefs to ________ reefs to atolls.

23. The ________ Zone was designed to proclaim sovereign rights over these resources.

24. Ocean ridges are formed by __________ forces and ______________ volcanic activity.

25. Ooze can be calcareous or ______________ in composition and is made of the microscopic _________ of plants and animals.

MULTIPLE CHOICE

1. The _________ is a submersible submarine which has contributed greatly to our understanding of the oceanic ridges.
   a. H.M.S. *Beagle*
   b. *Glomar Challenger*
   c. JOIDES *Resolution*
   d. *Alvin*

2. Which of the following is NOT part of an ophiolite sequence?
   a. pillow lavas
   b. layered gabbro
   c. massive rhyolite
   d. deep-sea sediments

3. The internal character of the oceanic crust beneath the sea floor has been determined by
   a. deep drilling.
   b. seismic profiling.
   c. study of ophiolites.
   d. all of the above

4. The steepest slopes of the continental margins occur on the
   a. continental shelf.
   b. continental rise.
   c. continental slope.
   d. abyssal plain.

5. Submarine canyons cut across the
   a. abyssal plains.
   b. ocean ridges.
   c. trenches.
   d. none of the above

6. Sediments attributed to turbidity currents are characterized by which sedimentary structure(s)?
   a. large-scale crossbedding
   b. mudcracks
   c. ripple marks
   d. graded-bedding

7. Turbidity currents deposit sediments on
   a. submarine fans and the abyssal plains.
   b. the abyssal plains and ridge crests.
   c. the continental shelf and continental slope.
   d. none of these

8. Which of the following is NOT a characteristic of a passive continental margin?
   a. wide continental shelf
   b. well developed continental rise
   c. deep sea trench
   d. close proximity of abyssal plains

9. Which of the following are associated with active continental margins?
   a. mountains
   b. volcanoes
   c. high seismic activity
   d. all of the above

10. The abyssal plains are flat because
    a. they were eroded flat during low sea level stand.
    b. they are covered with flat-lying sediment.
    c. they are capped by flat basalt flows.
    d. all of the above

11. Deep-sea trenches have
    a. low heat flow and low seismic activity.
    b. high heat flow and high seismic activity.
    c. low heat flow and high seismic activity.
    d. huge positive gravity anomalies.

12. Most of Earth's intermediate and deep focus earthquakes are associated with
    a. mid-ocean ridges.
    b. deep-sea trenches.
    c. abyssal plains.
    d. passive continental shelves.

13. A rift valley extends
    a. along an ocean ridge crest.
    b. along the floor of a deep-sea trench.
    c. across most abyssal plains.
    d. none of these

14. Which of the following is NOT associated with black smokers?
    a. high, rapidly building chimneys
    b. deep-focus earthquakes
    c. unusual life forms not dependent on sunlight
    d. deposits of metal sulfides

15. Sea floor fractures produce offsets in
    a. submarine canyons.
    b. trenches.
    c. guyots.
    d. ridge crests.

16. A seamount with a flat top is called a(n)
    a. guyot.
    b. atoll.
    c. abyssal hill.
    d. aseismic ridge.

17. The most abundant type of sediment in the deep sea is
    a. cosmic dust.
    b. gravel from rivers.
    c. pelagic clay and ooze.
    d. deposits resulting from reaction with sea water.

18. Ooze is sediment that
    a. is carried to the continental shelves by streams during periods of low sea level.
    b. sediment that is carried by icebergs then dropped when the ice melts.
    c. clay formed from the shells of planktonic organisms.
    d. sediment that is blown out to the middle of the ocean, then settles gradually to the bottom.

19. Reefs that are separated from an island by a lagoon are called
    a. patch or pinnacle reefs.
    b. atolls.
    c. barrier reefs.
    d. fringing reefs.

20. The Exclusive Economic Zone extends _____ nautical miles from the coast.
    a. 12
    b. 100
    c. 200
    d. 1000

21. Manganese nodules may prove to be a valuable source of ______ in the future.
   a. manganese
   b. cobalt
   c. nickel
   d. all of these

22. Hydrothermal vents are important because
   a. they support a diverse community of sealife
   b. they form ores of various metal sulfides
   c. they form through the burning of methane hydrates
   d. answers A and B only

23. Methane hydrate
   a. could supply much of our energy needs if we find a way to recover it.
   b. could drastically increase global temperatures.
   c. is found within many Exclusive Economic Zones.
   d. all of these answers

24. Surface ocean currents
   a. have only local effects.
   b. transfer deep water to the surface.
   c. transfer heat from the equator to the poles.
   d. all of these answers

25. Which ocean has more trench systems and more composite volcanoes?
   a. Chilean
   b. Pacific
   c. Arctic
   d. Atlantic

TRUE OR FALSE

____1. The ancient Greeks discovered most of the features of the sea floor almost 2000 years ago.

____2. Echo sounding is used to determine the internal geology of the oceanic crust.

____3. Gabbro is an important constituent of the oceanic crust.

____4. The edge of the continental crust does not necessarily correspond to the shorelines.

____5. The most gently sloping part of the continental margin is on the continental shelf.

____6. During the glacial advances of the Pleistocene, sea level was generally higher than today.

____7. All submarine canyons connect to major rivers mouths on the continents.

____8. Passive continental margins usually lack a continental rise.

____9. The west sides of South and North America represent passive continental margins.

____10. Passive continental margins typically occur within plates rather than at their margins.

____11. Along the bottom of an ocean basin, seawater has a temperature very close to zero degrees Centigrade.

___12. The deepest parts of the ocean occur in the abyssal plains.

___13. Gravity surveys indicate that trenches are not in isostatic equilibrium.

___14. Sea floor forms at a mid-ocean ridge.

___15. Iceland is a large segment of mid-ocean ridge sticking above the surface of the ocean.

___16. Organisms found associated with submarine hydrothermal vents at mid-ocean ridges are not dependent on photosynthesis at the base of the food chain.

___17. Aseismic ridges are characterized by intensive seismic activity.

___18. Most deep-sea sediment is finer-grained than sediment deposited on the margins of continents.

___19. Coral reefs require clear, warm, shallow water.

___20. Fringing reefs are attached to an island or continent.

___21. Oozes are deep-sea sediments rich in windblown dust and volcanic ash derived from the continents.

___22. Over 15% of U.S. oil production is from continental shelf areas.

___23. The EEZ was a region set aside as a sanctuary for marine life.

___24. Phosphorite, a source of phosphorous, forms in nodules on the deep abyssal plain.

___25. Atolls are evidence for plate motion.

DRAWINGS AND FIGURES

1. Draw a cross-section of the seafloor emphasizing the various layers of the oceanic crust. Use Fig. 12.6 to check your answer.

2. Sketch a series of 3 diagrams depicting the transition of an island from an active volcanic island with a fringing reef to an inactive volcanic island with a barrier reef to an atoll.

Chapter 13
Deformation, Mountain Building, and the Evolution of Continents

CHAPTER OBJECTIVES

By the end of this chapter you should be able to:

1. Differentiate between stress and strain and define the various categories of stress and strain.
2. Identify the factors that cause different styles to occur in a rock.
3. Describe the orientation of a tilted plane such as a bedding plane or a fault plane.
4. Name the parts of a fold and classify folds.
5. Differentiate between joints and faults and classify faults.
6. Describe several ways that mountains are formed.
7. Define orogeny and describe several processes that take place during an orogeny.
8. Compare orogeny along the various types of convergent margins and give examples of each.
9. Explain how continents grow by accretion and state the evidence for that process.
10. Differentiate between shields, platforms, and cratons.

USEFUL ANALOGIES

It is easy to look at a cross-section of a fault (e.g. those in question 3 in the *Drawings and Figures* section below) and figure out which block is the hanging wall and which is the footwall.

Draw a little vertically standing stick figure man across the fault so that the fault cuts him at the waist. If you draw it properly, the man's head will be in one block and his feet in the other. The block hanging over his head is the hanging wall and that in which his feet sit is the footwall.

There are many materials that you can use to demonstrate how various factors influence the type of strain that a material will show. One of the best is Silly Putty®.
When you pull the putty slowly it behaves in a plastic manner but when you snap it quickly it breaks. Therefore, the rate at which stress is applied is important in determining how Silly Putty® deforms.

A sheet of paper provides another example. Compress it and it folds (plastic behavior) but pull it and it tears (brittle behavior). Therefore, the direction of applied stress is important in determining how paper deforms. However, it doesn't matter the rate at which you apply the stress to paper because it will behave the same way regardless.

KEY TERMS

After reading this chapter, you should be familiar with the following terms:

deformation
stress
strain
compression
tension
shear stress
elastic strain
plastic strain
fracture
brittle
ductile
strike
dip
geologic structure
folds
monocline
anticline
syncline
axial plane
limb
symmetrical fold
asymmetrical fold
overturned fold
recumbent fold
fold axis
plunging fold
dome
basin
joints
fault
fault plane
fault scarp
fault breccia
hanging wall block
footwall block
relative movement
dip-slip fault
normal fault
reverse fault
thrust fault
strike-slip fault
left-lateral
right-lateral
oblique-slip fault
geologic map
mountain
mountain range
mountain systems
block faulting
horst
graben
orogeny
Alpine-Himalayan orogenic belt
circum-Pacific orogenic belt
orogen
accretionary wedge
continental accretion
terrane
shield
platform
craton
Canadian Shield

CHAPTER CONCEPT QUESTIONS

1. What is the difference between stress and strain?

2. What is the difference between brittle and plastic strain? What factors determine if rocks will be plastic or brittle in their behavior? What geologic structures result from brittle vs. ductile behavior?

3. Explain how the orientation of a plane is described using strike and dip.

4. Contrast in terms of fold limbs a monocline, an anticline and a syncline.

5. Define, in terms of fold limbs, axes, and axial planes, the following types of folds: (a) symmetrical, (b) asymmetrical, (c) overturned, (d) recumbent, (e) nonplunging, (f) plunging.

6. How can you tell whether a recumbent fold is an anticline or a syncline?

7. Compare and contrast an anticline and a dome (or a syncline and a basin).

8. Compare and contrast a joint and a fault.

9. Define, in terms of hanging wall and footwall, (a) normal fault and (b) reverse fault.

10. Under what condition is a reverse fault considered a thrust fault?

11. What kinds of forces cause normal and reverse faulting?

12. How can you tell if a fault is a left-lateral or a right-lateral strike-slip fault?

13. Briefly describe four different ways that mountains form.

14. Compare and contrast orogeny along an oceanic-oceanic convergent boundary with an oceanic-continental convergent boundary and a continental-continental convergent boundary. Give an example of each.

15. What geologic evidence suggests that parts of some continents (the West Coast of North America, for example) were transported in from distant parts of Earth?

16. Why are highly deformed metamorphic rocks such a common part of the continental crust, even under the craton far from the plate margins?

COMPLETION QUESTIONS

1. Stress is ________ per unit area.

2. Squeezing rocks creates a ________ stress, pulling rocks in opposite directions creates ________ stress, and ________ stress develops when parallel forces act in opposite directions.

3. Rocks may behave as ________ materials to produce fractures, or they may deform by plastic strain to create ________.

4. The trend of a horizontal line in a dipping bedding plane is called its ________, while the angle of inclination between the bedding plane and a horizontal plane is called the ________.

5. An upward flexure of the rock layers forms a(n) ________, while a downward flexure creates a(n) ________.

6. The ________ divides a fold into two equal halves. Each half is called a ________.

7. If a fold axial plane is not vertical, the fold is said to be ________, if it is horizontal the fold is ________.

8. In a nonplunging fold, the fold axis is ________, while in a plunging fold the fold axis is ________ from the horizontal.

9. In a ________ the beds dip away from the center in all directions. In a ________ the beds dip toward the center from all directions.

10. In a basin or syncline the rocks exposed at the center of the structure are ________ in age than those exposed on the outer parts of the limbs.

11. If relative movement occurs between rocks on either side of a fracture, the fracture is said to be a ________. If no movement as occurred, the fracture is a ________.

12. In a normal fault the hanging wall moves ________ relative to the footwall.

13. In a reverse fault the ________ wall moves up relative to the ________.

14. In a thrust fault the relative motion is the same as in a ________ fault, but the angle of dip of the fault plane is less than ________ degrees.

15. ________ forces are responsible for strike-slip faults.

16. Oblique faults have both ________ and ________ displacement.

17. A fold with the oldest layers in the core (center) is an ____________________ .

18. In areas that are being stretched due to tensional stresses, the uplifted blocks, called ________, form the mountains; and the down-dropped blocks, called ________, form valleys.

19. An episode of deformation of the Earth's crust to form mountains is called an ________.

20. Most present-day orogenic activity occurs along two belts: the __________ and the ___________ Orogenic Belts.

21. Sediments and even a sliver of sea floor (ophiolite) may be scraped off a descending plate during subduction and form a highly deformed terrain called an ________.

22. The Andes Mountains are rising where a(n) ________-bearing plate is converging with a(n) ________-bearing plate.

23. The Himalayas are rising where a(n) ________-bearing plate is converging with a(n) ________-bearing plate.

24. Much of the West Coast of North America developed from accretion of ________, which differ from one another in terms of fossil content, paleomagnetism and structural trend.

25. Regions of exposed, ancient granitic and metamorphic crustal rocks are called ________ . Areas where these are buried beneath sedimentary rocks are called __________.

MULTIPLE CHOICE

1. If a rock permanently bends but does not break it is demonstrating
   a. elastic behavior.
   b. plastic behavior.
   c. brittle behavior.
   d. none of the above

2. Plastic versus brittle rock behavior is determined by
   a. the temperature.
   b. the rock type.
   c. the type of stress.
   d. all of the above

3. Strike may be defined as
   a. the angle from a horizontal plane down to a bedding plane.
   b. the intersection of a horizontal plane with a tilted plane.
   c. the angle of the dip of the fold axis.
   d. none of these

4. The number adjacent to a strike and dip symbol on a geologic map indicates
   a. the strike of a bed.
   b. the dip of a bed.
   c. neither a nor b
   d. both a and b

5. A fold with only one dipping limb is a(n)
   a. monocline.
   b. anticline.
   c. syncline.
   d. recumbent fold.

6. In an anticline
   a. the beds arch upwards.
   b. the limbs dip toward each other.
   c. the youngest rocks are in the center of the fold.
   d. all of the above

7. A(n) ________ connects points of maximum fold curvature in all thereby dividing the fold in half.
   a. fold axis.
   b. axial plane.
   c. limb.
   d. plunge.

8. If the axial plane is horizontal, the fold is classified as a
   a. recumbent fold.
   b. symmetrical fold.
   c. asymmetrical fold.
   d. plunging fold.

9. In a plunging fold the _________ is inclined from the horizontal
   a. strike
   b. axial plane
   c. fold axis
   d. none of the above

10. In a dome the beds
    a. dip away from a fold axis.
    b. dip toward the center in all directions.
    c. dip toward a fold axis.
    d. dip away from the center in all directions.

11. Joints are
    a. fractures with vertical displacement.
    b. fractures with horizontal offset.
    c. fractures with both vertical and horizontal offset.
    d. none of the above

12. Faults are fractures that can show
    a. vertical offset.
    b. horizontal offset.
    c. either a or b
    d. no displacement at all

13. The "hanging wall"
    a. overlies the fault plane.
    b. overlies the axial plane.
    c. lies beneath the fault plane.
    d. lies beneath the axial plane.

14. The hanging wall of a reverse fault
    a. moves down relative to the footwall.
    b. moves up relative to the footwall.
    c. moves horizontally away from the footwall.
    d. remains stationary.

15. If you are standing by a right-lateral strike-slip fault, looking across the fault plane the displacement of the opposite block appears to be
    a. up.
    b. down.
    c. horizontally to the left.
    d. horizontally to the right.

16. The San Andreas Fault in California is a
    a. large right-lateral strike-slip fault.
    b. large left-lateral strike-slip fault.
    c. large normal fault.
    d. large thrust fault.

17. Fault block mountains are due to
    a. thrust faulting.
    b. reverse faulting.
    c. strike-slip faulting.
    d. normal faulting.

18. Horsts form _________ while grabens form _________.
    a. valleys / mountains
    b. normal faults / reverse faults
    c. mountains / valleys
    d. none of the above

19. Thrust faults are very common in many mountain belts and indicate that the mountains resulted from ________ stress.
    a. compressional
    b. extensional
    c. shear
    d. none of the above

20. Which would you LEAST expect to find at an ocean-continent convergent margin?
    a. volcanic mountains
    b. thrust faulted mountains
    c. accretionary wedge
    d. You find ALL of them.

21. A continental-continental plate collision forms
    a. a trench.
    b. a mountain range.
    c. a chain of volcanic islands.
    d. all of the above

22. Which of the following may become a terrane?
    a. a mid-ocean ridge
    b. a seamount
    c. a small fragment of continental crust
    d. any of the above

23. The most common geologic structures are
    a. normal faults.
    b. folds.
    c. joints.
    d. monoclines.

24. Force per area is
    a. strain.
    b. deformation.
    c. structure.
    d. stress.

25. How do faults differ from joints?
    a. Joints form during ductile deformation.
    b. Rocks on either side of the joint are offset parallel to the fracture.
    c. Rocks on either side of the fault are offset parallel to the fracture.
    d. Joints form by brittle deformation.

TRUE OR FALSE

____1. The same material may behave in a ductile manner when under compression and a brittle manner under tension.

____2. Strain is deformation due to stress.

____3. If a material returns to its original shape once stress is released, it is exhibiting plastic strain.

____4. As rocks become more deeply buried they become more brittle.

____5. Dip is the angle of inclination between an imaginary horizontal plane and a bedding plane.

____6. A syncline is a fold where the two limbs dip toward one another.

____7. Overturned folds always have an upside down limb.

____8. In a basin the beds dip toward the center in all directions.

____9. Joints are the commonest structure in rocks

____10. A fault scarp is a permanent feature on the landscape and is the best way to locate faults.

____11. Unlike bedding planes, fault planes lack a strike and dip.

____12. In a normal fault, the hanging wall moves up relative to the footwall.

____13. Thrust faults are particularly shallow dipping reverse faults.

____14. Oblique-slip faults have both dip-slip and strike-slip offset.

____15. A mountain system comprises several mountain ranges.

____16. The Basin and Range Province formed by thrust faulting.

____17. Block-faulted mountains are due to tensional stresses in the crust.

____18. The Alpine-Himalayan belt and the circum-Pacific belt include most of the area on Earth currently under compression at converging plate boundaries.

____19. Most of Earth's largest mountain ranges resulted from tensional stresses at divergent plate boundaries.

____20. Accretionary wedges are largely made up of highly folded and faulted marine sedimentary rocks.

____21. Much of the West Coast of North America is composed of terranes that originated from different parts of the Earth.

____22. Shields are areas of craton that are currently tectonically unstable.

____23. Anticlines and synclines are rarely found together.

____24. Mountains always form at the edges of continents where the active margins are.

____25. Anticlines commonly form structural traps for oil.

DRAWINGS AND FIGURES

1. In the adjacent block sketch and label the axial planes then label all fold axes and limbs. Also label an anticline and which is a syncline.

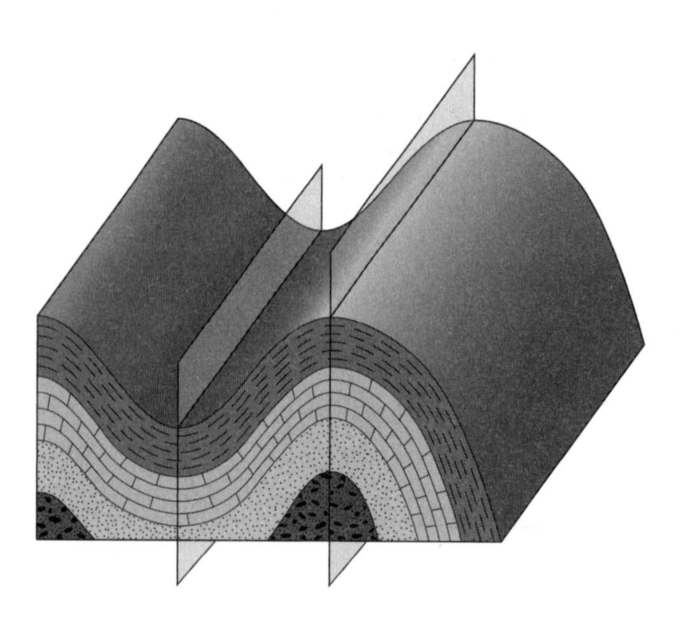

2. Shown below are five cross-sections of folds. Identify completely the type of each. Layers are numbered from oldest(1) to youngest(3).

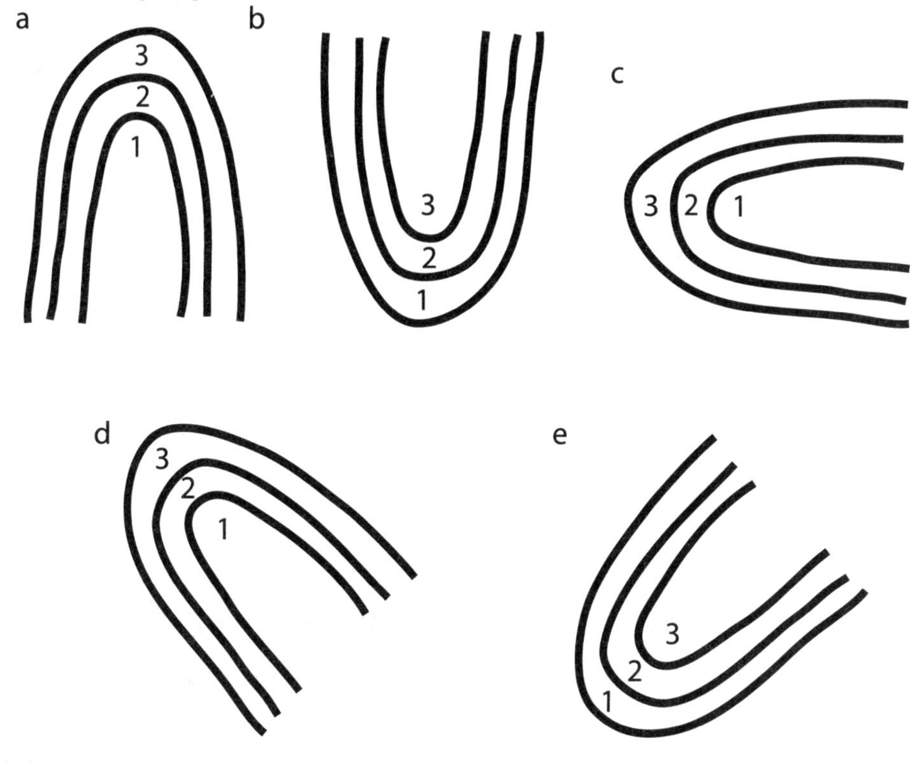

3. Shown below are two cross-sections of faults. Label the normal fault and the reverse fault.

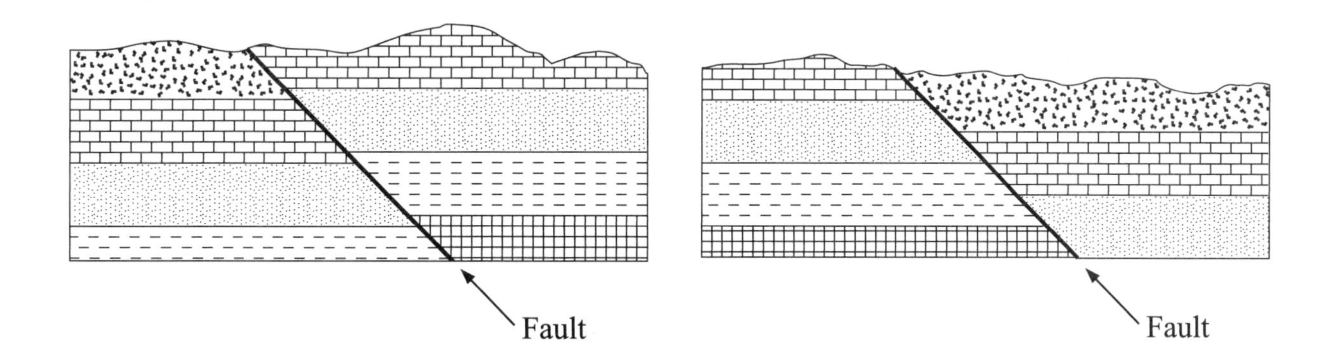

4. The figure below is a map view of a strike-slip fault that has offset some railroad tracks. Is it right or left-lateral?

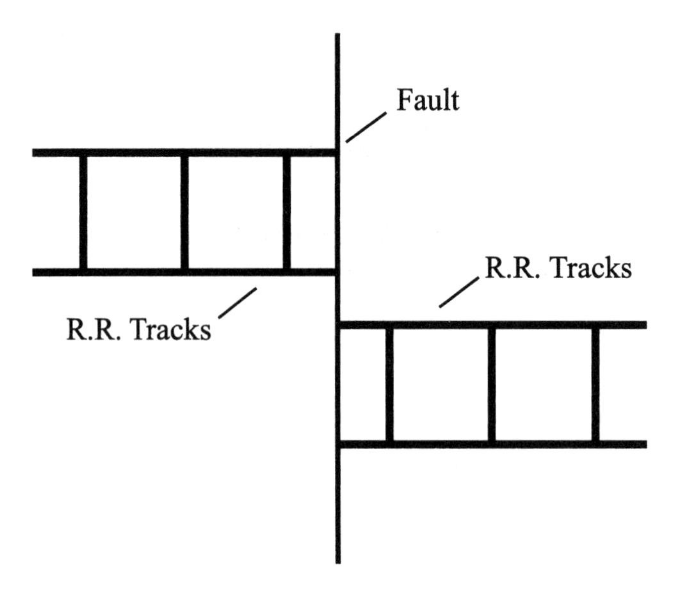

Chapter 14

Mass Wasting

CHAPTER OBJECTIVES

By the end of this chapter you should be able to:

1. Define mass wasting.
2. List the factors that define a slope's shear strength.
3. List and discuss the factors that influence an area's susceptibility to mass wasting.
4. List and differentiate between the different types of mass wasting.
5. Describe what sorts of activities a hazard assessment study entails.
6. Describe a number of ways the detrimental effects of mass wasting can be mitigated.

KEY TERMS

After reading this chapter, you should be familiar with the following terms:

landslide
mass wasting
shear strength
angle of repose
dynamic equilibrium
rapid mass movement
slow mass movement
rockfalls
talus
slide
slump
rock slide
mudflow
debris flow
earthflow
quick clays
solifluction
permafrost
creep
complex movement
debris avalanche
slope stability map
cut-and-fill
benching
rock bolts

CHAPTER CONCEPT QUESTIONS

1. What is mass wasting and why does it occur?

2. What factors define a slope's shear strength?

3. What is the significance of the angle of repose?

4. Describe how each of the following factors influences mass wasting: (a) slope angle, (b) weathering and climate, (c) water content, (d) overloading, (e) vegetation, and (f) bedrock geology of the area. For each of these, describe a situation in which man or nature may manipulate a particular factor so as to induce a mass movement.

5. Distinguish between rapid and slow movement. List examples of each.

6. Distinguish between the general categories of mass movement (i.e. falls, slides and flows).

7. Define rockfall and list three factors that can cause it.

8. Compare and contrast slumps and rock slides. What factors can lead to each?

9. Distinguish between a debris flow, a mudflow, and an earthflow.

10. What is solifluction and what factors lead to it?

11. Why is creep so often overlooked? What signs might tip you off that it is occurring?

12. What does "complex movement" mean? List a couple of examples.

13. When performing a hazard assessment, what signs of mass movement history and potential do geologists look for?

14. Describe three ways in which a hillside that has a high potential for mass movement can be stabilized.

COMPLETION QUESTIONS

1. A slope will be stable as long as its ________ strength is greater than the downslope pull of ________.

2. The steepest angle that most slopes in unconsolidated material can maintain without collapsing (called the angle of ________) is between ____ and ____ degrees for unconsolidated material.

3. A steep slope is generally ________ stable than a gentle slope.

4. Water acts to reduce the ________ between the soil particles.

5. A slope that is ________ to the dip of the underlying sedimentary bedrock layers may be particularly unstable.

6. The most common triggering mechanism for slope failure is the shaking due to ________ and excessive amounts of ________.

7. If the movement of material is fast enough to be visible, then it is considered to be ________ mass movement.

8. ________ piles constructed by rockfall accumulate at the base of cliffs.

9. In a ________ there is a downward movement and rotation of material along a curved surface of rupture.

10. Where a hill slope is parallel to the dip of the underlying sedimentary bedrock, ________ may occur.

11. A mass movement in which the material behaves in a viscous or plastic manner is termed a ________.

12. All slopes are in _______ equilibrium, which means if humans change them, they will ______ in response.

13. ________ flows are similar to mudflows, but are composed of larger-sized particles, contain less water and flow more slowly.

14. Earthflows move more ________ than mudflows or debris flows.

15. ________ clays spontaneously liquefy and flow when disturbed by shaking.

16. In the high latitudes and altitudes, the seasonal melting of the upper ________ can saturate soil and cause a mass movement known as ________.

17. The slowest and most common type of downslope movement is soil ________.

18. Movement of earth material that involves more than one type of mass movement is called ________ movement.

19. Information derived from a geologic hazard assessment of an area is often compiled on ________ maps as an aid to planners and engineers.

20. Many road cuts along interstate highways are stabilized by cutting a series of step-like terraces. This process is called ________.

21. Mass wasting is more likely if the weathering zone is __________, the slope is ______, and the rock has lots of __________ minerals.

22. Road construction frequently triggers mass wasting because excavation __________ the slope.

23. Water influences slope stability by decreasing _________and increasing the ___________.

24. Arid and semi-arid regions may have mass wasting because rainfall can occur in intense ________ and there is little ________ to buffer its effects.

25. Slopes that exceed the angle of repose will either ________ or are made of ________ rock.

MULTIPLE CHOICE

1. Which of the following would be considered a type of mass wasting event?
   a. slump
   b. rock slide
   c. debris avalanche
   d. all of the above

2. The most important stress opposing a slope's shear strength is imparted by
   a. running water.
   b. earthquakes.
   c. frost wedging.
   d. gravity.

3. A thick weathering zone
   a. does not affect a slope's stability.
   b. causes a slope to be more stable.
   c. causes a slope to be less stable.
   d. can not occur in regions of steep slopes.

4. Water can encourage mass flow by
   a. reducing friction between grains in a regolith.
   b. undercutting a steep slope.
   c. weathering bedrock to clay.
   d. all of the above

5. Removal of vegetation tends to make a slope
   a. more susceptible to mass flow.
   b. less susceptible to mass flow.
   c. steeper.
   d. increase its shear strength.

6. Overloading usually results from
   a. plate tectonics.
   b. undercutting of waves and streams.
   c. human activity.
   d. freezing and thawing.

7. The most stable geologic setting occurs when
   a. the beds dip into a hillside.
   b. the beds dip in the same direction as the slope of the ground.
   c. a dense fracture pattern is present.
   d. All of the above are equally stable.

8. Which of the following is considered to be a slow mass movement
   a. mudflow.
   b. debris flow.
   c. soil creep.
   d. rock fall.

9. Talus cones result from
   a. rock fall.
   b. rock slide.
   c. mud flows.
   d. any of the above

10. In a slump, the earth materials
    a. move forward by sliding down bedding or fracture plane.
    b. fall from a cliff in clumps.
    c. move downward and outward along a curving rupture plane.
    d. none of the above

11. In a rock slide materials
    a. move forward by sliding down bedding or fracture plane.
    b. fall from a cliff in clumps.
    c. move downward and outward along a curving rupture plane.
    d. none of the above

12. Mudflows
    a. are common in arid and semiarid areas.
    b. move very rapidly.
    c. comprise mostly silt and clay sized material.
    d. all of the above

13. Which of the following moves the slowest?
    a. mudflow
    b. debris flow
    c. earthflow
    d. They all move about the same speed.

14. One would expect solifluction to be a particularly important process
    a. where waves are undercutting sea cliffs.
    b. in high alpine meadows of mountains in Alaska.
    c. in areas of very sparse rainfall such as Death Valley, California.
    d. in a Brazilian rain forest.

15. Generally speaking, the most common form of mass movement is
    a. debris flow.
    b. slumping.
    c. mudflow.
    d. none of the above

16. Which of the following is NOT a sign of creep?
    a. curved tree trunks
    b. a steep scarp
    c. bulging retaining walls
    d. tilted telephone poles

17. The most common type of complex movement is a combination of
    a. fall and slide.
    b. fall and flow.
    c. slide and flow.
    d. none of the above

18. A debris avalanche is a complex movement combining
    a. fall and slide.
    b. fall and flow.
    c. either a or b
    d. neither a nor b

19. Which of the following types of mass movement is the most difficult to recognize in performing hazard assessments?
    a. rock slide
    b. soil creep
    c. slumping
    d. mudflow

20. One of the most effective ways to stabilize a slope is to
    a. remove all excess vegetation.
    b. increase the compaction of soil by adding water to the slope.
    c. control and remove the water going into the subsurface.
    d. none of these

21. Oversteepening can be caused by undercutting by
    a. streams.
    b. waves.
    c. road building.
    d. all of these answers

22. Of the following, the fastest and least viscous type of mass movement is
    a. earthflow.
    b. solifluction.
    c. mudflow.
    d. creep.

23. In which of the following would rockbolts be an effective landslide mitigation?
    a. water-saturated horizontal layers of clay
    b. a hillside of deeply weathered sandstone
    c. granite cut by fractures near the surface
    d. a sandstone cliff along the California coast undermined by wave erosion

24. Slow mass movements
    a. include solifluction.
    b. occur only on gentle slopes.
    c. occur only in loose debris.
    d. include debris avalanches.

25. Overloading of a slope
    a. occurs only through human action.
    b. increases the driving force of gravity by increasing the weight.
    c. reduces compaction.
    d. all of these answers

TRUE OR FALSE

___1. Mass wasting can only occur in steep terrain.

___2. Dry loose material cannot accumulate at a steeper angle than the angle of repose.

___3. Weathering of bedrock helps to stabilize the ground surface making it less susceptible to mass movement.

___4. Water may trigger slope failures by reducing the friction between soil particles.

___5. Vegetation tends to stabilize a slope by binding soil particles together.

___6. Loading a slope with large amounts of excavated material compacts the subsurface and increases the slope's stability.

___7. Earthquakes have triggered some of the Earth's most devastating mass wasting events.

___8. Rapid mass wasting means that the earth materials move an appreciable distance over the time span of a human life.

___9. Rapid mass wasting is responsible for transport of a significantly greater volume of material than slow mass wasting.

___10. Earthflows are rapid, devastating flows usually triggered by violent earthquakes.

___11. In a slide, material acts as a viscous fluid or a plastic.

___12. Under cutting by rivers can trigger rock slides.

____13. Mudflows are common in arid and semiarid regions where they are triggered by heavy rainstorms.

____14. A scarp occurs at the top of a slump.

____15. Solifluction can occur in all climates, but it is most common in the tropics.

____16. Permafrost makes a good firm base on which a structures can be built.

____17. Soil creep is one of the slowest downslope movements.

____18. Curved tree trunks may indicate that creep is taking place.

____19. Once a landslide has occurred in an area, it generally stabilizes and is a good building site.

____20. The occurrence of subsurface clay layers tends to make a steep slope more stable.

____21. A good drainage system on a hillside is an effective, yet inexpensive, way to lessen a slopes susceptibility to mass wasting.

____22. Rock bolts may prove to be effective at preventing rock slides.

____23. Slopes at angle of repose are always stable.

____24. Flows often begin as falls, slumps, or slides.

____25. Benching is frequently used to reduce the mass wasting hazards along roadcuts.

DRAWINGS AND FIGURES

1. Identify the type of mass flow illustrated in the following block diagrams.

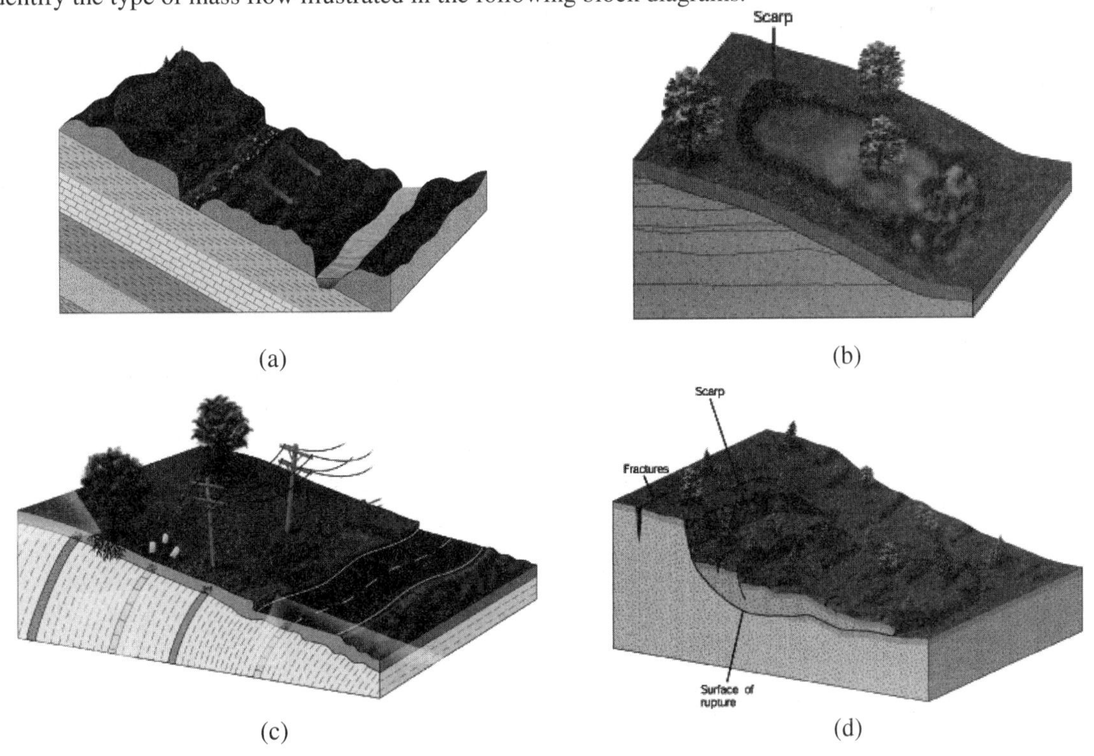

(a) (b) (c) (d)

2. What will be the result of this roadcut?

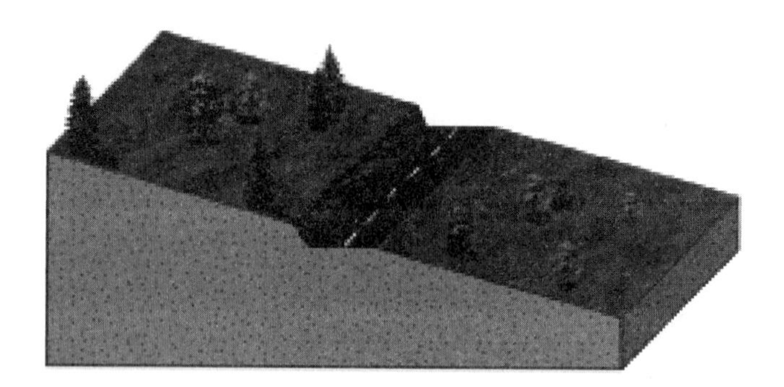

Chapter 15

Running Water

CHAPTER OBJECTIVES

By the end of this chapter you should be able to:

1. Say how Earth's water is distributed.
2. Reproduce the hydrologic cycle and discuss its parts.
3. Differentiate between laminar and turbulent flow.
4. Define infiltration capacity and discuss how it determines whether precipitation will result in surface water or groundwater.
5. Differentiate between sheet and channel flow.
6. Describe how gradient, velocity and discharge are determined and discuss what factors influence their values.
7. Describe how a stream erodes and what processes occur during erosion.
8. Differentiate between the different types sediment loads that streams transport.
9. Compare and contrast braided streams and their deposits with meandering streams and their deposits.
10. Describe how delta deposition takes place and how various factors such as waves and tides can affect a delta.
11. Understand how stream discharge data is used in forecasting chances of flooding over different periods of time and how the data is used in flood prevention planning.
12. Describe the various types of drainage patterns and for each state what factors lead to its development.
13. Explain how the tendency of a stream to seek base level and become graded affects its erosion and deposition.
14. Describe the various processes that are important in valley development.
15. Discuss how superposed streams, stream terraces, and incised meanders form.

KEY TERMS

After reading the chapter, you should be familiar with the following terms:

hydrologic cycle
evaporation
condensation
precipitation
runoff
transpiration
laminar flow
turbulent flow
infiltration capacity
sheet flow
sheet erosion
channel flow
stream gradient
stream velocity
channel roughness
discharge
potential energy
kinetic energy
hydraulic action
abrasion
potholes
dissolved load

suspended load
bed load
saltation
competence
capacity
alluvium
braided streams
meandering streams
cut bank
point bar
ox-bow lake
flood plain
lateral accretion
natural levees
vertical accretion
topset beds
delta
prograde
bottomset beds
foreset beds
topset beds
distributary channels
stream-dominated delta
bird's foot delta
wave-dominated delta
tide-dominated delta
alluvial fan
hydrograph
flood-frequency curve
recurrence interval
drainage basin
divide
drainage patterns,
dendritic drainage
rectangular drainage
trellis drainage
radial drainage
deranged drainage
ultimate base level
temporary base level
longitudinal profile
graded stream
valley
gully
canyon
gorge
downcutting
lateral erosion
headward erosion
stream piracy
stream terrace
incised meander
superposed stream
water gap

CHAPTER CONCEPT QUESTIONS

1. Compare and contrast evaporation, condensation, transpiration, and precipitation.

2. How does laminar flow differ from turbulent flow? What conditions cause one to dominate over the other?

3. Explain infiltration capacity. What characteristics of a soil influence infiltration capacity?

4. How is gradient calculated?

5. Explain how gradient, channel shape and channel roughness affects stream velocity.

6. State three reasons why the velocity of a river typically increases in a downstream direction even though the gradient decreases.

7. What is discharge and how is it calculated?

8. How do the concepts of potential and kinetic energy relate to a stream?

9. List and describe the different mechanisms by which streams carry sediment of different sizes.

10. What is the difference between capacity and competency?

11. What factors would cause a stream to become a braided stream?

12. How do meandering streams migrate? How do oxbow lakes form?

13. How does velocity vary in a meandering river? How does it influence where erosion and deposition take place?

14. What is meant by "lateral accretion" vs. "vertical accretion"? Where does each of these occur?

15. How does a delta form? How do tides and waves modify the shape of a delta?

16. How do alluvial fans form? Why are they most abundant in arid areas?

17. Comment on the usefulness and accuracy of flood-frequency curves in predicting floods over the short term (year to year) and the long term (decade to decade). What part do they have in planning flood control programs?

18. What sorts of structures have been built to control floods?

19. What is meant by a river's drainage basin? Give five examples of ways geology influence a drainage pattern.

20. What is base level? What happens to a stream if base level is lowered? How might a stream's base level be raised?

21. How does an ungraded stream modify itself to become at least partly graded?

22. Describe three types of erosion that contribute to valley development?

23. What is stream piracy and how does it occur?

24. Explain how each of the following forms: (a) stream terraces, (b) incised meanders, and (c) superposed streams.

25. Detail the stages leading to the creation of a "mature" stream (such as the Mississippi River) according to the classic model. What are some shortcomings of this model?

COMPLETION QUESTIONS

1. Approximately ____ % of the water on Earth is in the ocean and another ____ % is tied up in glaciers.

2. Although only ____ % of the evaporation on Earth is from continents, about ____ % of the precipitation falling to Earth falls on land.

3. ________ flow is characterized by a mixing of stream lines.

4. The amount of rain that will become runoff is determined by the ground's ________, the maximum rate at which water can be absorbed.

5. When runoff begins, water flows in a thin film called ________ flow. Sooner or later it becomes confined in long trough-like depression called a ________.

6. The ________ of a stream is calculated by dividing its vertical drop by the horizontal distance that it has traveled.

7. The velocity of a stream is determined dividing the horizontal distance a stream by a unit of ________.

8. Deep meandering channels cut into bedrock are called ________ .

9. Channel roughness influences flow velocity because of the ________ between the particles on the bottom of the channel and the flowing water.

10. ________ is the total volume of water in a stream moving past a particular point in a given period of time.

11. ________ streams cut right through resistant geologic structures.

12. Sedimentary particles are set in motion by the ________ action of the flowing water.

13. Potholes in solid bedrock in the beds of streams are formed by ________.

14. Streams carry sand and gravel as ________ load, silt and clay particles as ________ load, and ions in solution as ________ load.

15. ________ refers to the maximum-sized particles a stream can carry, and ________ refers to the total amount of sediment a stream can carry.

16. Braided streams have diverging and rejoining ________, and form if the stream is carrying excessive ________.

17. Meanders can be abandoned, forming ________ lakes.

18. The outside curve of a meander, called the ________, is a site of erosion while the inside of the curve is a site of deposition where a ________ is formed.

19. When streams overflow their banks, fine-grained sediments are deposited on the ________.

20. Stream terraces form by a lowering of ________, which causes renewed downward erosion.

21. A delta forms where a sediment-laden stream enters a ________ body of water, its velocity ________ and its load is deposited.

22. In a stream dominated delta, the delta ________ far seaward.

23. The diverging streams in a delta are called ________.

24. The Mississippi River Delta is a ________ dominated delta with a ________ foot shape.

25. An ________ forms where a sediment laden stream exits the mouth of a canyon and its load is quickly deposited on the valley floor.

26. Discharge data collected at gauging stations are used to construct a ________ for a stream, which is a plot of stream discharge over ________.

27. The entire area from which a stream receives its discharge is called the ________ basin. Drainage ________ form the boundaries of this area.

28. A common drainage pattern resembling the veins in the leaf of a tree is called ________.

29. A ________ pattern is characteristic of a stream system flowing off a volcanic peak.

30. A rectangular drainage pattern may develop if the bedrock in the area has a pattern of right angle intersecting ________.

31. The lowest level to which a stream can erode its channel is ________.

32. The ultimate base level for all the world's streams is ____ _______.

33. ________ streams have achieved a delicate balance between gradient, discharge, flow velocity, and channel characteristics such that neither significant erosion nor deposition occurs in the stream channel.

34. Valleys are made deeper by stream ________ and made longer by ________.

MULTIPLE CHOICE

1. Not including water frozen in glaciers, fresh water comprises ________ % of the water on Earth.
   a. over 70
   b. about 50
   c. about 10
   d. less than 1

2. ______ dominates in all natural streams.
   a. Laminar flow.
   b. Turbulent flow.
   c. Both a and b occur equally.
   d. Neither a nor b are present in natural streams.

3. Which of the following has the highest infiltration capacity?
   a. a clay-rich soil which has been compacted by herds of cattle
   b. water saturated clay soil
   c. dry, loose gravel
   d. a tar coated parking lot

4. Compute the average gradient of a 3 km long stream, which has a source at an elevation of 3000 meters above sea level and is 600 meters above sea level at its mouth.
   a. The gradient can not be computed due to insufficient information.
   b. 3000 m/km
   c. 800 m/km
   d. 800 km/m

5. The velocity of a stream is controlled by the
   a. discharge.
   b. channel shape.
   c. channel roughness.
   d. all of the above

6. A stream flows fastest in a channel that is
   a. deep and narrow.
   b. shallow and wide.
   c. semicircular in cross-section.
   d. Channel shape is not a factor in stream velocity.

7. A stream has a cross-sectional area of 100 square meters and is flowing at 2 meters/second. what is its discharge?
   a. 200 cubic meters/second
   b. 50 cubic meters/second
   c. 200 square meters/second
   d. 50 square meters/second

8. Which of the following types of stream load spends the most time in contact with the bottom?
   a. dissolved load
   b. suspended load
   c. bed load
   d. They all spend the same amount of time.

9. Capacity is
   a. the total amount of water a stream is carrying.
   b. the maximum-sized particle stream can carry.
   c. the total amount of load a stream can carry.
   d. none of the above

10. Braided streams are characterized by
    a. excessive bed load.
    b. point bars.
    c. oxbow lakes.
    d. all of the above

11. Meanders can
    a. become deeply incised in narrow canyons.
    b. migrate back and forth across a valley.
    c. be cut off to form ox-bow lakes.
    d. all of the above

12. A natural levee is located between
    a. the channel and floodplain.
    b. a delta and the oceans.
    c. a point bar and the channel.
    d. an alluvial fan and mountain front.

13. Deltas which prograde the farthest into a lake or sea are usually
    a. stream-dominated.
    b. tide-dominated.
    c. wave-dominated.
    d. alluvial fan dominated.

14. Stream channels traversing the top of a delta are called
    a. tributaries.
    b. distributaries.
    c. secondary channels.
    d. braided channels.

15. The marshy areas between delta distributaries are commonly sites of deposition of sediment that eventually lithifies to become
    a. conglomerate.
    b. limestone.
    c. coal.
    d. dolostone.

16. The steeply dipping front of a delta is formed from
    a. distributaries.
    b. topset beds.
    c. bottomset beds.
    d. foreset beds.

17. Flood-frequency curves have proven most useful in
    a. predicting precisely when a flood will occur over the next ten years.
    b. predicting how high the water will rise for a given amount of rainfall.
    c. long-term planning flood control programs.
    d. predicting where along a stream flooding will most frequently occur.

18. The area from which a stream and its tributaries receive their discharge is called the
    a. drainage basin.
    b. drainage divide.
    c. flood plain.
    d. alluvial fan.

19. Dendritic drainage patterns develop where
    a. many intersecting fractures occur.
    b. high, isolated mountain peaks exists.
    c. where alternating resistant and nonresistant tilted beds occur.
    d. none of the above

20. Recently glaciated areas in which the drainage has had no time to organize commonly exhibit ________ drainage.
    a. radial
    b. trellis
    c. dendritic
    d. deranged

21. The ultimate base level for streams in general is
    a. sea level.
    b. approximately 150 meters above sea level.
    c. the bottom of the deepest abyssal plains.
    d. none of the above

22. Stream valleys are lengthened by
    a. headward erosion.
    b. lateral accretion.
    c. downcutting.
    d. sheet runoff.

23. Stream terraces form
    a. by downward erosion due to a rise in base level.
    b. by downward erosion due to a lowering of base level.
    c. where the velocity decreases and the stream load is deposited.
    d. none of the above

24. In general, incised meanders form
    a. by an uplift of the land, and downward erosion.
    b. by headward erosion.
    c. where a stream's velocity decreases and its load is deposited.
    d. where extensive lateral erosion can occur.

25. Braided streams are most common in ____________ environments.
    a. humid
    b. glacial
    c. desert
    d. marsh

TRUE OR FALSE

____1. Most of the Earth's fresh water is in rivers and streams.

____2. Although oceans make up seventy percent of the Earth's surface, over half the precipitation falls on land.

____3. Water flows because it possesses no strength.

____4. In laminar flow, water particles move in parallel paths.

____5. Runoff occurs when infiltration capacity is exceeded.

____6. The most important source of water in stable, steadily flowing streams is groundwater.

____7. Discharge is defined as the total amount of sediment carried by a stream.

____8. The steep mountain tributaries to large rivers generally flow faster than the large rivers themselves.

____9. Water stored behind a dam is an example of kinetic energy.

____10. If running water contains only dissolved load, but no solid sedimentary particles, little or no abrasion of the bedrock over which it flows will occur.

____11. Silt and clay are transported as bed load.

____12. Sediment transported by saltation spends much of the time on the bottom.

____13. Braided streams are common on alluvial fans and in streams fed by glacial meltwater.

____14. As a stream rounds a bend, there is erosion on the inside and deposition on the outside of a bend.

____15. A point bar develops when a meander is cut off.

____16. Coarse sand and gravel typically dominate vertically accreted floodplain sediments.

____17. Distributary channels deposit a delta's topset beds.

____18. A bird's foot delta is a stream-dominated delta.

____19. Alluvial fans develop best in arid and semiarid regions.

____20. Mudflows are common features on alluvial fans.

____21. Flood-frequency curves are extremely accurate in predicting where and when floods will occur.

____22. Folded sedimentary rocks typically force a trellis drainage pattern on the landscape.

____23. Radial drainage forms in a region with numerous fractures.

____24. A stream cannot erode below its base level.

____25. Superposed streams can cut right across resistant bedrock ridges.

___26. In the evolution of a stream valley according to the classic model, the deepening of the valley by downcutting is followed later by the lateral erosion and valley widening

DRAWING AND FIGURES

1. Draw a diagram of the hydrologic cycle. Include in your diagram condensation, precipitation to ocean and land, evaporation from ocean and land, transpiration, runoff, and groundwater. Fig. 15.4 may give you some ideas, although your version does not need to be so fancy.

2. Using the adjacent graph, answer the following questions.
   a. How fast must a current be flowing to erode
      sand 1/16 mm in diameter?
      clay 1/256 mm in diameter?
      gravel 2 mm in diameter?

   b. As a stream slows down, at what velocity will each of the above sediments be deposited?

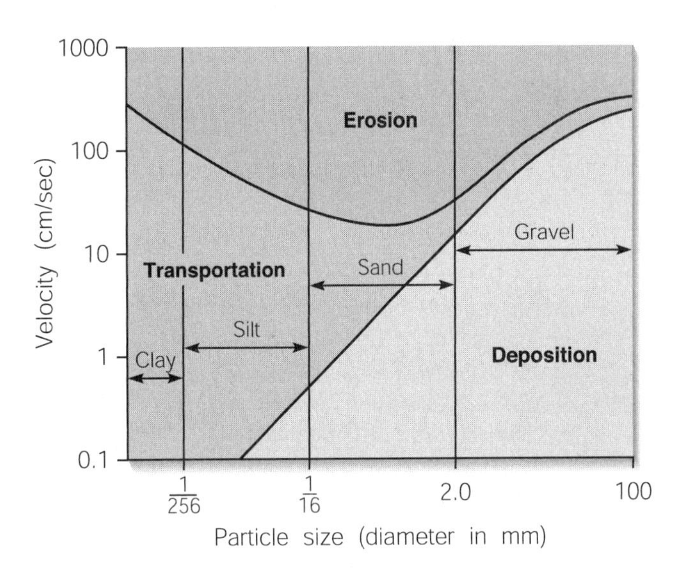

3. In the adjacent diagram, show by an arrow the path of the fastest flow. Label where erosion and deposition are taking place.

4. According to the flood-frequency curve below, over what period of time would you expect a stream to experience a discharge of
   a. 75m^3/sec.
   b. 150m^3/sec, and c) 350 m3/sec

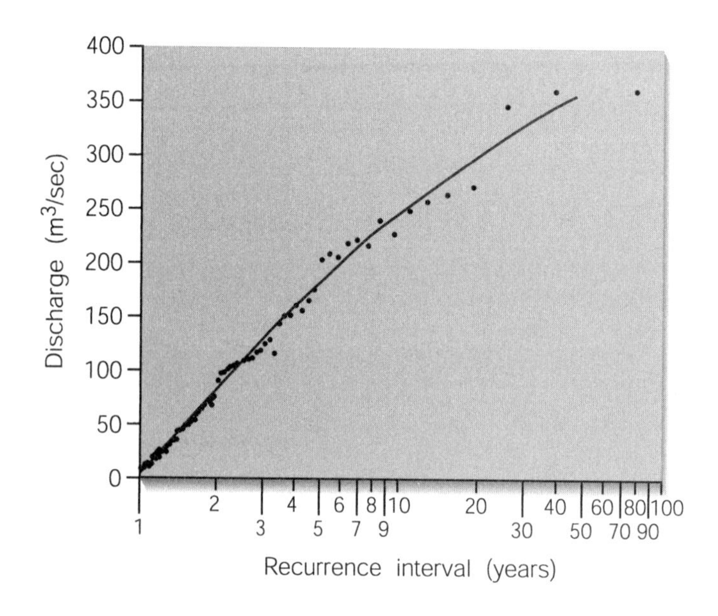

5. Label the type of drainage patterns illustrated by each of the block diagrams below.

6. Identify the feature pointed out in the adjacent photograph. Figure 15.17b in the textbook is a similar photograph.

Chapter 16

Groundwater

CHAPTER OBJECTIVES

By the end of this chapter you should be able to:

1. Explain the difference between porosity and permeability and describe factors that influence porosity and permeability of a sediment or sedimentary rock.
2. Describe the relationship between the water table, the capillary fringe, the zone of aeration, and the zone of saturation.
3. Compare groundwater movement in the zone of aeration with that in the zone of saturation.
4. Explain what springs are and what they indicate.
5. Explain what artesian wells are and the condition that result in them.
6. Describe the characteristics and origin of karst terrain.
7. Explain how cave systems develop and how this development is related to the water table.
8. Describe a number of cave formations.
9. Discuss what happens when a well is drilled and some of the consequences of water production in excess of recharge.
10. Describe several sources of groundwater contamination.
11. Differentiate between hot springs and geysers.
12. Discuss how geothermal energy is produced, its merits and problems.

USEFUL ANALOGIES

To illustrate the differences between porosity and permeability compare a block of Styrofoam with a sponge. You can easily see that both have high porosity because each has as much air in its total volume as solid mass. However, if you allow water to drip on both only the sponge will absorb water. In fact, Styrofoam is used to make cups! You don't see many sponge cups! Styrofoam, therefore, has low permeability.

A sponge also is good for illustrating how water can move against gravity by capillary pull. If you barely touch a sponge to the surface of a puddle of water it will flow upward. This is what happens to groundwater at the capillary fringe. Capillary action, high porosity and high permeability are what make sponges useful.

KEY TERMS

After reading this chapter, you should be familiar with the following terms.

groundwater
porosity
permeability
aquifer
aquiclude
suspended water

zone of aeration
zone of saturation
water table
capillary fringe
recharge
spring
perched water table
water well
cone of depression
artesian system
artesian-pressure surface
sinkhole
solution valley
karst topography
disappearing streams
cave
cavern
dripstone
stalactites
stalagmites
column
drip curtain
travertine terrace
saltwater incursion
cone of ascension
subsidence
hard water
hydrothermal
fumaroles
hot springs
mud pot
geyser
travertine (calcareous tufa)
siliceous sinter (geyserite)
geothermal energy

CHAPTER CONCEPT QUESTIONS

1. Why do you think the study of groundwater has emerged as a high priority of geologists in recent years?

2. How do porosity and permeability differ? Why are they significant?

3. What is the difference between an aquifer and an aquiclude? What characteristics of a rock formation would cause it to be one or the other? What types of rocks make good aquifers and good aquicludes?

4. How can normally non-porous rocks such as basalt or most limestone become effective aquifers?

5. Compare and contrast the movement of groundwater in the zone of aeration, the zone of saturation, and the capillary fringe. What drives the water movement in each?

6. What causes an aquifer to be discharged? List several ways in which an aquifer may be naturally recharged.

7. What do springs indicate? Why do springs sometimes occur at different elevations, even in a small area?

8. What causes a producing water well to go dry?

9. What is meant when a well is described as being artesian? What geologic conditions cause an artesian well?

10. What are the characteristics of an area characterized by karst topography? What geologic and climatic conditions cause karst topography?

11. Describe two ways in which sinkholes form.

12. Describe the evolution of a cave and its speleothems (the stalactites, stalagmites, etc.).

13. What is the High Plains aquifer? Why is it important and how is it endangered?

14. What geologic conditions make an area susceptible to saltwater incursion? What human actions cause it to occur? What solutions to the problem have been suggested?

15. What causes subsidence to occur as a result of groundwater pumping? Give two examples of where this has occurred.

16. List four sources of ground water pollution.

17. Where and how do geysers and hot springs form? What conditions cause a hot spring to be a geyser?

18. How is hot groundwater harnessed for energy? What are some advantages and disadvantages of this energy source? Name two places in the world where geothermal energy is being used today.

COMPLETION QUESTIONS

1. Groundwater makes up about ________ % of the world's supply of fresh water.

2. ________ is the percentage of space in a body of sediment or rock and ________ is the ability of that material to transmit fluid.

3. For a material to be permeable it must have ________ pore spaces.

4. Groundwater is transmitted in a permeable layer called an ________ and kept in that layer by surrounding impermeable layers called ________.

5. The ________ separates the zone of aeration from the zone of saturation.

6. At the water table is a very narrow zone called the ________ ________ where the water is transmitted upward.

7. In the zone of aeration groundwater flows downward driven by ________.

8. In the zone of saturation groundwater flows laterally from regions of ________ pressure to regions of ________ pressure.

9. A spring forms where the ________ ________ intersects the surface of the ground.

10. A ________ is formed when a local aquiclude is present in the zone of aeration.

11. A spring may form above the regional water table if the ground surface intersects a ________ aquifer.

12. To be effective a water well should be drilled to the zone of ________.

13. Pumping of groundwater creates a cone of ________ around the well.

14. For artesian conditions to exist an aquifer must be confined above and below by ________, the rock sequence must be ________ so that the aquifer formation is exposed at a higher elevation in the recharge area, and there must be sufficient ________ in the recharge area to keep the aquifer filled.

15. In an artesian well, pressure pushes the water up toward the ________ surface.

16. Most groundwater erosion is by a weak acid formed from the combination of water and ________ from the atmosphere.

17. Karst topography develops in a terrain underlain by ________ bedrock.

18. A ________ results when a cave collapses.

19. Most of the dissolution of a cave takes place in the zone of ________.

20. In a cave ________ form on the ceiling and ________ form on the floor, and if they join they form a ________.

21. The High Plains aquifer accounts for approximately ___ percent of the water used for irrigation in the U.S.

22. In coastal areas undergoing rapid development, removal of excessive amounts of fresh groundwater is resulting in ________ __________.

23. Reduction in water pressure in the pore spaces of sediments is resulting in compaction of the sediments and ________ of the ground surface in large areas of the arid western U.S.

24. Careless handling of sewage, landfills, and toxic waste disposal can result in ________ of the groundwater.

25. To be considered a hot spring the temperature of the water must exceed ___ degrees C.

26. The heat for most hot springs and geysers comes from cooling ________ beneath the surface.

27. Deposits of ________ often collect around the vents of geysers.

28. In a geyser the water is forced out of the ground by expanding ________.

29. In some areas of the world like California, Iceland, and New Zealand, ________ energy is used to generate electricity or heat homes.

MULTIPLE CHOICE

1. The largest use of groundwater is ________, which accounts for 65% of groundwater produced.
   a. household use
   b. heavy industry
   c. irrigation
   d. geothermal energy

2. Groundwater occurs in
   a. the pore spaces between clasts of sand or gravel.
   b. in fractures.
   c. aquifers.
   d. all of the above

3. Porosity is
   a. the ability to transmit fluids.
   b. the amount of void space between grains.
   c. the amount of water pressure present.
   d. none of the above

4. Permeability is
   a. the ability to transmit fluids.
   b. the amount of void space between grains.
   c. the amount of water pressure present.
   d. none of the above

5. A good aquiclude should have ________ porosity and ________ permeability.
   a. high / high
   b. high / low
   c. low / high
   d. low / low

6. The water table marks the
   a. top of the zone of aeration.
   b. the bottom of the zone of saturation.
   c. the top of the zone of saturation.
   d. none of the above

7. The capillary fringe is located
   a. at the water table.
   b. below the zone of saturation.
   c. above the zone of aeration.
   d. none of the above

8. In the zone of saturation groundwater moves
   a. from regions of high pressure to regions of low pressure.
   b. downward toward a lake or stream.
   c. both a and b
   d. neither a nor b

9. A spring is
   a. a man made opening in the zone of saturation.
   b. found where the zone of saturation intersects the ground surface.
   c. where the zone of aeration meets the ground surface.
   d. none of the above

10. A spring located at an elevation above the water table probably indicates
    a. excessive production of wells in the area.
    b. the presence of a perched aquifer.
    c. collapsing caves below the ground.
    d. a problem with saltwater incursion.

11. When water is pumped out of the ground, it creates a _____ around the well.
    a. cone of ascension
    b. cone of depression
    c. cone of incursion
    d. cone of recharge

12. Artesian conditions require
    a. confinement of the aquifer between aquicludes.
    b. a recharge area that is lower in elevation than the area in which water is being produced.
    c. arid conditions in the recharge area.
    d. all of the above

13. In an artesian system the artesian-pressure surface is always located
    a. above the producing aquifer.
    b. below the producing aquifer.
    c. above the ground.
    d. at sea level.

14. Karst topography results from
    a. the incursion of salt water in coastal areas.
    b. the existence of hundreds of artesian wells along a mountain front.
    c. the dissolution of limestone bedrock by groundwater.
    d. all of the above

15. Stalactites and stalagmites form when a cave is
    a. just beginning to dissolve.
    b. in the zone of aeration.
    c. in the zone of saturation.
    d. beneath the zone of saturation.

16. A vertical sheet of rock hanging from a fracture in the ceiling of a cave is
    a. a stalactite.
    b. a stalagmite.
    c. a travertine terrace.
    d. a drip curtain.

17. The effects of salt water incursion have been effectively combated by
    a. pumping fresh water back into the aquifer.
    b. increasing the rate of pumping of the producing wells.
    c. pumping "salt-cleaning" chemicals into the aquifer.
    d. All of the above have been successful.

18. Over pumping of groundwater causes
    a. lowering of the water table.
    b. saltwater incursion.
    c. ground subsidence.
    d. all of the above

19. Which of the following is the best place for a septic tank?
    a. in the zone of aeration in material of high permeability
    b. in the zone of aeration in material of low permeability
    c. in the zone of saturation in material of high permeability
    d. in the zone of saturation in material of low permeability

20. The temperature of a hot spring is
    a. above the boiling point.
    b. above that of the human body.
    c. above the average air temperature of a region.
    d. above 10°C.

21. Siliceous material deposited around the vent of a geyser is called
    a. travertine terrace.
    b. travertine tufa.
    c. sinter.
    d. none of the above

22. Geothermal energy makes up about ___ % of the energy harnessed by humans.
    a. 2
    b. 10
    c. 50
    d. 75

23. Which of the following substances would be the most porous?
    a. massive limestone
    b. well-sorted sandstone
    c. unfractured granite
    d. gneiss

24. Groundwater erodes by
    a. hydraulic pressure.
    b. hydrostatic pressure.
    c. chemical reactions.
    d. recharge reactions.

25. Carlsbad Caverns is in the desert of southwestern New Mexico. It
    a. indicates New Mexico used to get more rainfall and developed karst topography.
    b. shows karst can form in any climate with limestone at the surface.
    c. is small (but beautiful) since it is in the desert.
    d. connects to Mammoth Cave, Kentucky.

TRUE OR FALSE

____1. Groundwater commonly occurs as underground rivers.

____2. A well sorted sediment usually has a higher porosity than a poorly sorted sediment.

____3. If a material is porous, it will be permeable too.

____4. An effective aquiclude has low permeability.

____5. In the zone of saturation the pore spaces are completely filled with water.

____6. Except in arid areas, the water table rises and falls approximately parallel to the overlying surface topography.

____7. The water well is an artificial opening in the zone of saturation.

____8. To have a water well pump water all year long, the well must be in the zone of aeration all year long.

____9. Artesian systems are highly desirable because they cannot be overproduced.

____10. Artesian systems are highly desirable because they typically contain much more pure water than other types of groundwater.

____11. Low concentrations of sulfuric acid in rain water are responsible for most cave formation.

____12. Karst topography is restricted to regions of arid climate.

____13. Sinkholes form by an explosive eruption of carbon dioxide gas released by the dissolving of limestone.

____14. Dripstone deposits are composed of calcium carbonate.

____15. Some of America's most productive agricultural land is in danger because of overproduction of groundwater.

____16. Because salt water is so much less dense than fresh water, deepening a well can mitigate salt water incursion in coastal aquifers.

____17. Unlike surface water, groundwater is generally impervious to industrial pollution.

____18. In the United States, hot springs are most abundant west of the Mississippi River.

____19. Geothermal water can be very corrosive.

____20. Geysers form as a solidified magma forces groundwater to the surface.

____21. A primary attraction of geothermal energy is that it cannot be exhausted.

____22. Hot springs form only in volcanic areas.

____23. Desert oases are often formed by artesian springs.

____24. The water table is the same in humid and arid areas since its location depends chiefly on the Laws of Physics.

____25. The formation of karst topography requires the action of groundwater on soluble rocks at or near the surface.

DRAWINGS AND DIAGRAMS

1. On the cross-section face of the block diagram below, draw a line representing the water table. Label the zones of saturation and aeration. Also label the perched water table.

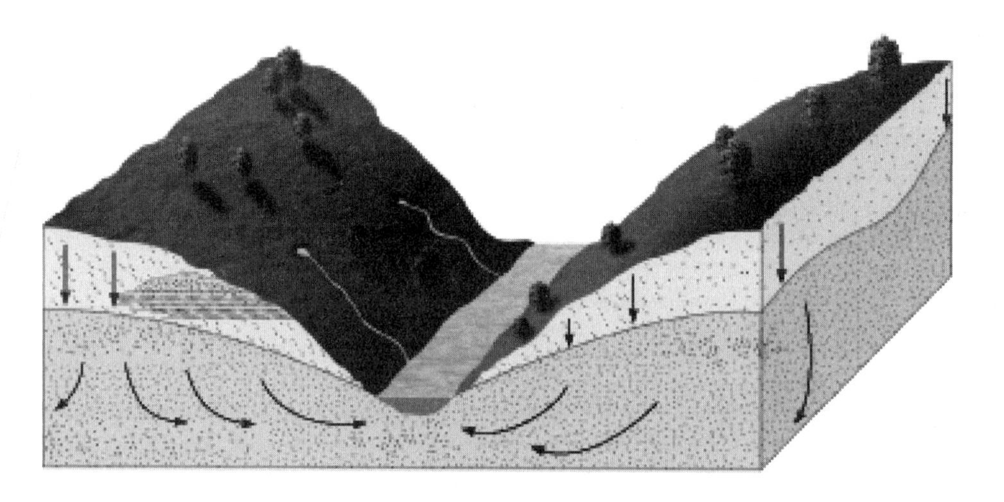

2. Sketch the interior of a cave with stalagmites, stalactites, soda straws and drip curtains.

Chapter 17

Glaciers and Glaciation

CHAPTER OBJECTIVES

By the end of this chapter you should be able to:

1. Distinguish between a glacier and other masses of snow and ice.
2. Describe how glacial ice originates and how it moves.
3. Distinguish between a valley glacier and a continental glacier.
4. Explain the concept of glacial budget and what determines if a glacier will advance or retreat.
5. Describe how rates of glacial movement vary within a glacier and with the seasons.
6. List and describe the various erosional processes associated with glaciers.
7. List, describe and recognize the various erosional and depositional landforms associated with both continental and valley glaciers.
8. Describe how the climate was different during the Pleistocene than it is today and what effects that different climate had on the landscape in different areas, particularly the arid western U.S. and coastal areas.
9. Discuss several theories for the cause of the Pleistocene Ice Ages.

USEFUL ANALOGIES

Have you ever left an ice tray in the freezer for a long period of time, then wondered why the cubes were so small when you finally wanted some ice? That's sublimation!

Lots of solids flow plastically given enough time. Old window glass panes often will be wavy and even may be thin at the top and thick at the bottom. This is because they flow, even more slowly than glacial ice.

KEY TERMS

After reading this chapter, you should be familiar with the following terms:

glacier
calving
sublimation
firn
glacial ice
plastic flow
basal slip
valley glacier (or mountain glacier, or alpine glacier)
piedmont glacier
continental glacier (or ice sheet)
ice cap
glacial budget
zone of accumulation
zone of wastage
firn limit
terminus
stagnant glacier
crevasse
ice fall
outlet glacier
glacial surge
outburst floods

bulldozing
plucking
roche moutonnée
abrasion
glacial polish
glacial striations
rock flour
U-shaped glacial trough
truncated spurs
fiord
hanging valley
cirque
tarn
arête
horn
ice-scoured plain
glacial drift
glacial erratics
till
stratified drift
end moraine
ground moraine
recessional moraine
terminal moraine
lateral moraine
medial moraine
drumlin
drumlin field
outwash plain
valley train
kettle
pitted outwash plain
kame
esker
varve
dropstone
Pleistocene
pluvial lake
proglacial lake
Milankovitch Theory

CHAPTER CONCEPT QUESTIONS

1. What is the difference between a glacier and a simple patch of mountain snow or ice? Why is an iceberg not a type of glacier?

2. How does glacial ice form? What makes it different from other ice, for example that on a frozen lake?

3. What causes a glacier to move? Describe the two mechanisms by which glaciers move.

4. Compare and contrast a valley glacier with a continental glacier, a piedmont glacier, and an ice cap.

5. Explain in terms of glacial budget and zones of accumulation and wastage, why a glacier advances or retreats.

6. How do flow rates vary within a single glacier?

7. Where, how and why do crevasses form?

8. What are glacial surges and to what are they attributed?

9. What is the difference between plucking and abrasion? What are the results of each?

10. How do you tell a valley that has been cut by a stream from one which has been carved by a glacier?

11. Explain the origin of the following: (a) fiord, (b) hanging valley, (c) arête, (d) cirque, and (e) horn.

12. Describe a landscape, such as that of Canada, which has been subjected to the scouring of a large continental glacier.

13. Contrast the origin and characteristics of stratified drift and till.

14. Compare and contrast recessional moraines, terminal moraines, lateral moraines and medial moraines. Which are restricted to valley glaciers?

15. Compare and contrast the origin of outwash plains, valley trains, kames and eskers.

16. Contrast the origin of the following types of glacial lakes: (a) tarns, (b) kettles, (c) pluvial lakes, and (d) proglacial lakes. Characterize the sedimentary deposit of a glacial lake.

17. Certain regions of Earth experienced notable differences in climate during the Pleistocene compared to today, even outside the polar regions. Describe some of these.

18. How did the glaciation effect the coastlines of the continents?

19. What is meant by isostatic rebound? How does it relate to Pleistocene glaciation? How and why has it varied in different parts of the North American Continent?

20. Describe the three parameters that vary to cause glaciations according to the Milankovitch theory.

21. What theories have been proposed for causes of short-term cold periods such as the Little Ice Age?

COMPLETION QUESTIONS

1. A glacier is a mass of ice composed of compacted and recrystallized _______ that flows under its own weight across the land.

2. Today glaciers cover about _______ of the surface of the land, they occur on all continents except _______.

3. The largest present day glaciers occur in _______ and _______.

4. About _______ % of the world's water is contained in glaciers.

5. Icebergs are masses of floating ice that enter the sea by breaking off a glacier in a process called _______.

6. Ice can evaporate without melting in a process called _______.

7. As snow accumulates, it is _______ by the weight of the overlying snow, refreezes and forms a granular type of ice called _______ which further compacts to form glacial ice.

8. _______ flow, permanent deformation of the ice, is the most important way that glaciers move.

9. Meltwater can reduce the _______ between the ice and the ground beneath allowing movement by the process of _______.

10. _______ glaciers are confined to high mountain valleys.

11. In Greenland and Antarctica, continental glaciers are more than _______ meters thick.

12. A glacier has two major zones. Yearly snow additions exceed losses in the zone of _______ but losses exceed additions in the zone of _______. The _______ limit separates the two.

13. Accumulation is due to _______ fall, and wastage is due to _______, _______, and the calving of icebergs.

14. If wastage exceeds accumulation, a glacier will _______, but if accumulation exceeds wastage, a glacier will _______; and if wastage and accumulation are in balance, the glacier will be _______.

15. In a valley glacier, the highest velocity is in the zone of ______.

16. The upper part of a glacier behaves as a brittle solid, and this causes ________ to develop when a glacier moves over irregularities in the valley floor.

17. During a glacial _______ glaciers may advance several kilometers in a year or less.

18. _______ occurs when glacial ice freezes in the cracks and crevices of a bedrock projection and pulls it loose. This results in a bedrock landform called a _______.

19. Abrasion by a glacier results in a smooth glacial _______ which is often marred by straight, shallow scratches called _______ .

20. The milky appearance of streams containing meltwater is generally due to rock _______.

21. Glacially eroded valleys are characterized by having a _______ shaped cross section.

22. As glaciers melt and sea level rises, glacial troughs may be flooded by the sea resulting in _______

23. A bowl-shaped depression at the head of a glacier is called a(n) _______, and the knife-like ridge separating two of these is called a(n) _______.

24. A sharp mountain peak surrounded on all sides by cirques is a_______.

25. Areas eroded by _______ glaciers tend to be rather flat, monotonous and characterized by a _______ drainage pattern.

26. All sediment deposited by a glacier is called glacial _______.

27. Glacial _______ are large boulders transported many tens to hundreds of kilometers from their points of origin.

28. If glacial sediment was deposited directly by ice leaving a poorly-sorted, disorganized deposit it is called _______. If however, streams sorted the deposit it is _______ drift.

29. A ridge of till deposited at a glacier's terminus is called a(n) _______ moraine. If it marks the point of furthest advance it is called a(n) _______ moraine.

30. If a minor glacial advance deposits an end moraine during an overall recession it is a _______ moraine.

31. Where two lateral moraines merge, a _______ moraine forms.

32. When a continental glacier overrides ground moraine it often shapes the till into elongate hills called _______.

33. Meltwater laden with sediment from a melting glacier forms deposits called _______ plains and _______ trains.

34. Depressions that form from a block of ice left in the outwash produces a _______.

35. Long sinuous ridges of stratified drift deposited by streams flowing under a glacier are called _______.

36. Stratified drift deposited in depressions in the surface of the ice and subsequently lowered to the ground when the ice melts are called _______.

37. Large stones deposited in otherwise very fine-grained lake sediment are called _______ and were probably carried into the lake by _______.

38. According to the _______ theory, glacial-interglacial episodes of the Pleistocene may be due to variations in the eccentricity of the Earth's orbit, the tilt of the axis and the precession of the equinoxes.

39. _______ major ice advances have been recognized in North America while Europe had six or seven. However at least _______ warm-cold cycles can be detected in deep sea cores.

40. Since the Pleistocene continental glaciers have melted, the land has been rising. This process is called glacial _______.

41. The Great Salt Lake is remnant of a once larger _______ lake called Lake Bonneville.

42. Proglacial lakes are impounded by the _______ one side and a _______ on the other.

43. The Great lakes presently contain _____% of the water contained in the Earth's freshwater lakes.

44. When the Pleistocene glaciation was at its maximum, sea level was _______ meters lower than it is today.

45. If the present glaciers melted, sea level would _______ by _______ meters.

46. Under the weight of the Pleistocene glacial ice, the crust of the Earth was depressed up to _______ meters below its position today.

MULTIPLE CHOICE

1. Which of the following is NOT a necessary condition for a mass of ice to be considered a glacier?
   a. It must advance in the winter more than it melts in the summer.
   b. It must originate by compaction and recrystallization of snow.
   c. It must flow under its own weight.
   d. It must be on land.

2. During the transformation from snow to glacial ice the percentage of air
   a. decreases.
   b. increases.
   c. is unaffected.

3. When a glacier sublimates it
   a. melts.
   b. evaporates without melting.
   c. travels forward faster than normally.
   d. forms icebergs.

4. The greatest volume of ice on Earth today is held in _______ glaciers.
   a. valley
   b. continental
   c. piedmont
   d. All of the above are about equal.

5. During the Pleistocene, continental glaciers covered large parts of
   a. North America and South America.
   b. North America, South America and Europe.
   c. North America, Europe and Asia.
   d. North America, Asia and Africa.

6. During a warming period velocity due to basal slip ______ and velocity due to plastic flow ______.
   a. decreases / increases.
   b. increases / decreases.
   c. increases / stays the same.
   d. stays the same / decreases.

7. When wastage exceeds accumulation, a glacier will
   a. retreat.
   b. advance.
   c. remain stationary.
   d. stop flowing.

8. Crevasses result from the _____ behavior of ice in the top forty meters of a glacier.
   a. brittle
   b. plastic
   c. elastic
   d. juvenile

9. Glacial abrasion produces
   a. rock flour.
   b. glacial polish.
   c. glacial striations.
   d. all of these

10. Which is NOT characteristic of a glacial trough?
    a. a broad, flat valley floor
    b. steep, often vertical valley walls
    c. V-shaped cross-section
    d. truncated spurs

11. A hanging valley forms when the base of the main valley glacier is topographically _____ the tributary glacier.
    a. higher than
    b. lower than
    c. the same as
    d. either **a** or **b**

12. Which of the following pair are both erosional in origin?
    a. cirques and moraines
    b. horns and cirques
    c. cirques and drumlins
    d. moraines and drumlins

13. Glacial till is
    a. unstratified drift.
    b. stratified drift.
    c. both a and b
    d. neither a nor b

14. Medial moraines form when
    a. two terminal moraines merge.
    b. two lateral moraines merge.
    c. an end moraine is over-ridden by a glacier.
    d. a mountain peak is surrounded on all sides by glaciers.

15. Which of the following is likely to show poor sorting and no stratification?
    a. an esker
    b. an outwash plain
    c. a drumlin
    d. a kame

16. Eskers form
    a. on top of the ice.
    b. in front of the ice.
    c. beneath the ice.
    d. none of the above

17. A recessional moraine is a type of
    a. end moraine.
    b. lateral moraine.
    c. medial moraine.
    d. ground moraine.

18. Each dark/light couplet in a sequence of varves represents
    a. one year.
    b. about 10 years.
    c. about 1000 years.
    d. about 1000000 years.

19. During periods of glacial advances the Sahara Desert was covered by
    a. glaciers.
    b. forests.
    c. a shallow sea.
    d. volcanoes.

20. The Great Lakes started out as
    a. kettles.
    b. proglacial lakes.
    c. pluvial lakes.
    d. tarns.

21. Pluvial lakes formed in the far western U.S. during the Pleistocene glaciation as a result of
    a. blocks of ice melting in outwash plains.
    b. impoundment of water by terminal moraines.
    c. water filling depressions carved in bedrock by glaciers.
    d. an overall cooler and wetter climate in the area.

22. When large continental glaciers advanced across the northern hemisphere during the Pleistocene
    a. the crust beneath the subsided.
    b. worldwide sea level fell.
    c. temperate, subtropical and tropical belts were compressed toward the equator.
    d. all of the above

23. The best explanation for short periods of glacial advance that span only a few centuries includes
    a. plate tectonics.
    b. Milankovitch cycles.
    c. frequent volcanic eruptions coupled with variations in solar energy.
    d. all of the above

24. The firn limit
    a. is the boundary between the zones of accumulation and wastage.
    b. is easily seen in the Summer.
    c. usually changes from year to year.
    d. all of these answers

25. Hazards associated with glaciers include
    a. extreme erosion episodes.
    b. glacial surges.
    c. glacial outburst floods.
    d. sudden advances.

26. Which of the following is NOT typical of a valley glacier?
    a. moraine
    b. drumlin
    c. outwash plain
    d. till

TRUE OR FALSE

____1. The most common type of glacier is frozen sea ice.

____2. Africa is the only continent that lacks glaciers.

____3. In the western U.S., glaciers are an important source of fresh water during the dry season.

____4. Ice is a mineral

____5. Glaciers are made of firn.

____6. During most of the year basal slip is the main mechanism by which glaciers move.

____7. Continental glaciers are unconfined by topography.

____8. The main valley glacier moves faster than its tributaries.

____9. If a glacier's budget is perfectly balanced, it is stagnant.

____10. Glacial striations are parallel to the direction of glacial flow.

____11. Glacial troughs and stream cut valleys cannot be distinguished by shape of profile alone.

____12. During the glacial advances of the Pleistocene Epoch, sea level was higher than it is today.

____13. Fiords are glacial troughs occupied by the sea.

____14. Horns are deposited by meltwater beneath a glacier.

___15. Arêtes are bowl-shaped depressions near the head of a glacier.

___16. Continental glaciers carve high jagged mountains forming cirques, horns, and arêtes.

___17. Stratified drift is called till.

___18. Medial moraines form at the end of a valley glacier.

___19. An end moraine forms during a time of balanced glacial budget, when a glacier neither advances nor retreats.

___20. Drumlins can be used to determine the direction of movement of a continental glacier.

___21. Most outwash streams are meandering rivers.

___22. The dark layer of a varve is deposited in the winter.

___23. The scientists of ancient Greece and Rome first suggested the presence of widespread glaciers during the Pleistocene.

___24. During the Pleistocene, the world's temperature was close to freezing, even near the equator.

___25. The Great Salt Lake is a remnant of a great proglacial lake that existed during the last glacial advance.

___26. Plate tectonics may play a part in causing glaciations by moving continents to polar areas.

DRAWINGS AND FIGURES

1. On the block diagram below, label (a) a hanging valley, (b) a horn, (c) an arête, (d) a U-shaped glacial trough, (e) a cirque, and (f) truncated spurs.

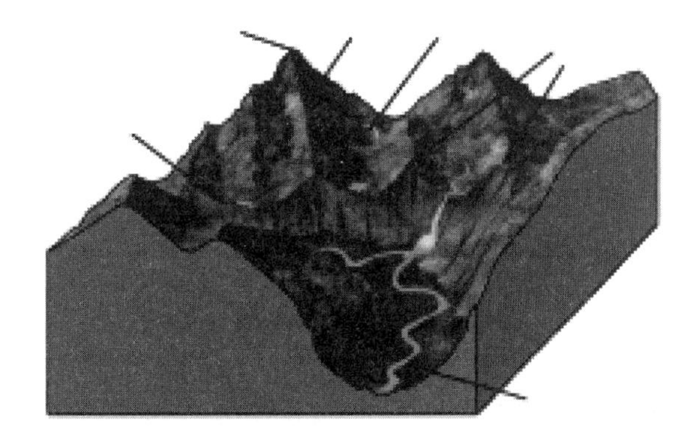

2. On the block diagram below, label (a) a kame, (b) drumlins, (c) the outwash plain, (d) kettle lakes, (e) an esker and (f) the end moraine.

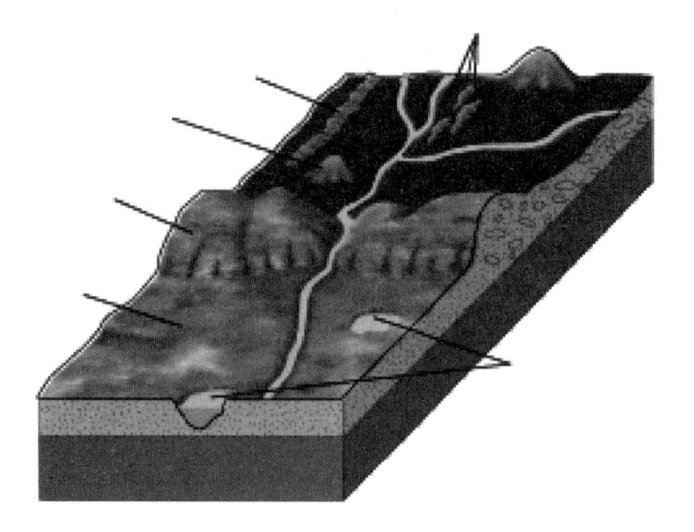

Chapter 18

The Work of Wind and Deserts

CHAPTER OBJECTIVES

By the end of this chapter you should be able to:

1. Describe how clasts of various sizes are moved by wind.
2. Compare and contrast water and wind as transporting agents.
3. Describe the main processes and products of wind erosion.
4. Define deflation and describe some of the products of that process.
5. Describe how dunes form and migrate.
6. Classify and describe the major dune types.
7. Define loess, and discuss its origin, character and distribution.
8. Describe what factors determine the distribution of deserts.
9. Describe the climatic conditions that constitute a desert and discuss how desert vegetation has adapted to these conditions.
10. Describe the important weathering processes of deserts and their products.
11. Describe how the surface and groundwater conditions in deserts differ from other areas.
12. Characterize, recognize and discuss the origin of desert landforms.

KEY TERMS

After reading this chapter, you should be familiar with the following terms:

desertification
suspended load
bed load
saltation
abrasion
ventifact
yardang
deflation
deflation hollow (or blowout)
desert pavement
dune
barchan dune
longitudinal dune (or seif dune)
transverse dunes
barchanoid dunes
parabolic dunes
star dune
loess
Coriolis effect
desert
rainshadow desert
rock varnish
internal drainage
playa lakes
playa or salt pan
alluvial fan
bajada
pediment
inselberg
mesa
butte

CHAPTER CONCEPT QUESTIONS

1. Distinguish between the bed load and the suspension load. What sizes are included in each in wind transport? Compare this to transport by water.

2. Why are sand-sized particles generally entrained by wind before finer silt and clay?

3. How does wind abrade rocks? Describe two products of abrasion commonly found in deserts.

4. What is deflation? Describe two products of deflation commonly found in deserts.

5. How do sand dunes form? How do they migrate?

6. What are the main factors that influence the size, shape, and arrangement of dunes?

7. List the four major dune types. For each describe (a) the orientation relative to the prevailing wind, (b) the conditions (i.e. sand supply, vegetation) under which that type forms, and (c) the typical range of sizes encountered.

8. What is loess? Where does it come from? Why is it important to agriculture?

9. What is the Coriolis effect, what causes it, and why is it significant in influencing wind direction?

10. What is the difference between a semi-arid and an arid region?

11. What is the difference between a low to mid-latitude desert and a rainshadow desert?

12. Characterize a desert in terms of temperature range, amount and nature of precipitation, soil development, and vegetation.

13. List at least two reasons why water erosion is so prominent in deserts despite the low rainfall.

14. How does the drainage pattern in deserts differ from that in more humid areas?

15. What are the major characteristics of a playa lake and salt pan?

16. Define and describe the origin of the following desert landforms: (a) alluvial fan, (b) bajada, (c) pediment, (d) inselberg, (e) mesa, and (f) butte.

COMPLETION QUESTIONS

1. Wind transports fine-grained sediments as ________ load and sand-particles as ________ load.

2. Abrasion causes a sandblasting effect that can form wind-faceted rocks called _______.

3. The removal of loose sediment by the wind is called _______.

4. Deflation may leave a blown out depression in the surface called a _______ or a mosaic of close-fitting pebbles left behind called _______.

5. Most sand dunes have an asymmetric profile with a gentle _______ slope and a steep _______ slope.

6. The steepest slope of a dry sand dune is between ____ and ____ degrees, the angle of _______.

7. A crescent-shaped dune with its tips pointing downwind is called a ________ dune, while a crescent-shaped dune with its horn pointing upwind is called a ________ dune.

8. A long, linear dune perpendicular to the wind is called a ________ dune, and a long, linear dune parallel to the prevailing wind direction is called a ________ dune.

9. _______ is fine-grained, wind transported material derived from deserts, glacial outwash, and floodplains in semi-arid areas.

10. Loess covers _______ percent of the Earth's land surface and _______ percent of the land surface in the United States.

11. Because of the _______ effect, winds are deflected to the _______ of their direction of motion in the Northern Hemisphere and to the _______ in the Southern Hemisphere.

12. Air that rises at the equator tends to cool and sink in belts between ____ and ____ degrees north and south latitude.

13. The Great Basin Region of the western U.S. is largely a vast _______ desert formed on the _______ side of the Sierra Nevada mountain range.

14. _______ weathering dominates in a desert, the main processes being _______ fluctuations and _______ wedging.

15. Rock varnish is a coating composed of iron _______ and _______ oxides.

16. Desert valleys in the southwestern United States commonly have _______ drainage resulting in the formation of _______ lakes.

17. An alluvial _______ forms at the mouth of a _______ where a stream deposits its load.

18. When two or more alluvial fans coalesce, a _______ can form.

19. Sloping surfaces extending from the base of a mountain are called _______.

20. Isolated remnants of mountains in desert areas are called _______.

21. _______ are broad, flat topped erosional remnants bounded on all sides by steep slopes. A smaller version of this is called a _______.

22. Desert soil is usually ________, present only in ______ if at all, and easily eroded because of the sparse vegetation.

23. Pediments get __________, as the mountains adjacent to them are _______ or eroded.

24. Desertification is a major problem in parts of _________, the _________ __________, Asia, and the United States.

25. Dunes are classified based on ______________ and their relation to wind ________ into four major types.

MULTIPLE CHOICE

1. Wind-blown sand is generally transported by
   a. saltation.
   b. suspension.
   c. either a or b
   d. neither a nor b

2. Wind generally entrains silt and clay-sized grains _______ sand-sized grains.
   a. before
   b. after
   c. about the same time as

3. Abrasion by wind commonly takes the form of
   a. pitting.
   b. etching.
   c. polishing.
   d. all of the above

4. Deflation can only occur if
   a. the sediments are loose.
   b. the sediments are anchored by vegetation.
   c. if the sediments are wet.
   d. all of the above

5. The steep side of an asymmetric sand dune
   a. faces upwind.
   b. faces downwind.
   c. can face either upwind or downwind.
   d. can not be used to determine the wind direction.

6. The tallest type of dune is
   a. transverse.
   b. parabolic.
   c. barchan.
   d. star.

7. Barchan dunes are
   a. crescent-shaped with tips pointing downwind.
   b. crescent-shaped with tips pointing upwind.
   c. linear dunes oriented parallel to the prevailing wind.
   d. linear dunes oriented perpendicular to the prevailing wind.

8. Transverse dunes are
   a. crescent-shaped with tips pointing downwind.
   b. crescent-shaped with tips pointing upwind.
   c. linear dunes oriented parallel to the prevailing wind.
   d. linear dunes oriented perpendicular to the prevailing wind.

9. Ancient sand dunes, now preserved as sandstone, would be characterized by an abundance of which sedimentary structure?
   a. scour and fill
   b. mud cracks
   c. graded bedding
   d. cross-stratification

10. Loess is dominated by
    a. silt and clay.
    b. sand and silt.
    c. sand and gravel.
    d. gravel.

11. The fertile soils of the Great Plains, the Midwest, and in the Mississippi Valley were derived from extensive
    a. sand dunes.
    b. loess deposits.
    c. playa lakes.
    d. alluvial fans.

12. In the northern hemisphere the Coriolis effect causes winds to be deflected
    a. to the right of their motion.
    b. to the left of their motion.
    c. always to the east, regardless of the direction of their motion.
    d. always to the west, regardless of the direction of their motion.

13. In high pressure zones
    a. hot, moist air rises.
    b. hot, moist air descends.
    c. dry, cool air rises.
    d. dry, cool air descends.

14. Deserts receive less than _______ centimeters of rain per year.
    a. 15
    b. 2.5
    c. 35
    d. 25

15. Most of the world's deserts are found in belts
    a. along the equator.
    b. in the low to middle latitudes.
    c. in the middle to high latitudes.
    d. above the Arctic Circle.

16. Where a high mountain range causes a rainshadow desert, the desert forms on _____ side(s) of the mountains.
    a. upwind (windward)
    b. downwind (leeward)
    c. both

17. Which is NOT an adaptation of plants of desert plants?
    a. wide spacing
    b. soft stems and leaves
    c. slow growth
    d. the ability to become dormant for long periods

18. Rock varnish contains oxides of _____ and iron.
    a. copper
    b. tin
    c. magnesium
    d. none of the above

19. Most deserts are dominated by a surface consisting of
    a. a cover of sand dunes.
    b. exposures of bedrock and desert pavement.
    c. loess deposits.
    d. playa lakes.

20. Alluvial fans coalesce to form
    a. playas.
    b. pediments.
    c. bajadas.
    d. buttes.

21. Star dunes are
    a. common along coasts.
    b. also called seif dunes.
    c. among the tallest dunes in the world.
    d. all of these answers

22. Among the best places to drill for groundwater in deserts are
    a. yardangs.
    b. bajadas.
    c. playas.
    d. ventifacts.

23. Water erosion is a powerful force in deserts because
    a. vegetation is sparse.
    b. rainfall is often in powerful cloudbursts.
    c. the valleys have internal drainage.
    d. answers A and B only

24. As population expands into semi-arid and arid areas, people increase the amount of dust in the air by
    a. removing desert plants.
    b. removing desert pavement.
    c. exposing land to deflation.
    d. all of these answers

25. Desertification
    a. is often accelerated by grazing too many animals in the region.
    b. is only a problem in Africa.
    c. can be reversed fairly easily.
    d. occurs faster naturally than people can do it.

TRUE OR FALSE

____1. Wind is a turbulent fluid.

____2. Wind is the most important erosional agent in deserts.

____3. Sand is commonly lifted up to ten meters by wind.

____4. In saltation, sand grains move in a bouncing fashion.

____5. The suspension load of wind may be carried thousands of kilometers.

___6. Sand usually travels by saltation up the leeward side of a dune, then avalanches down the windward side.

___7. Ventifacts form by deflation.

___8. Barchan dunes have their steep side facing downwind, and their tips pointing upwind.

___9. Parabolic dunes form where there is a partial cover of vegetation.

___10. Longitudinal dunes require large sediment supplies to form.

___11. Accumulations of loess deposits are a common feature of deserts.

___12. Loess deposits can cause big problems for agriculture because of their low fertility.

___13. In the Southern Hemisphere, the Coriolis effect deflects objects to the left of their intended path.

___14. Deserts form in belts of high pressure at the low to middle latitudes.

___15. Deserts have high temperatures throughout the year.

___16. By definition deserts have no plants living in them.

___17. Mechanical weathering generally dominates over chemical weathering in deserts.

___18. Evaporite minerals may be common in playa lakes.

___19. It is not uncommon for some deserts to receive their full year's rainfall in one cloudburst.

___20. Buttes and mesas form by streams eroding through a resistant cap rock to erode the less resistant rock below.

___21. Pediments are small erosional remnants of mountains.

___22. The groundwater table in deserts is higher under permanent streams.

___23. The type of dune formed depends on sand supply, wind direction and velocity, and amount of vegetation.

___24. Running water is responsible for most erosional landforms in deserts.

___25. Desert pavement forms by wind transport and deposition of large pebbles.

FIGURES AND DRAWINGS

1. Draw and explain how wind moves sand by saltation.

2. Identify the type of dunes illustrated by the five block diagrams below.

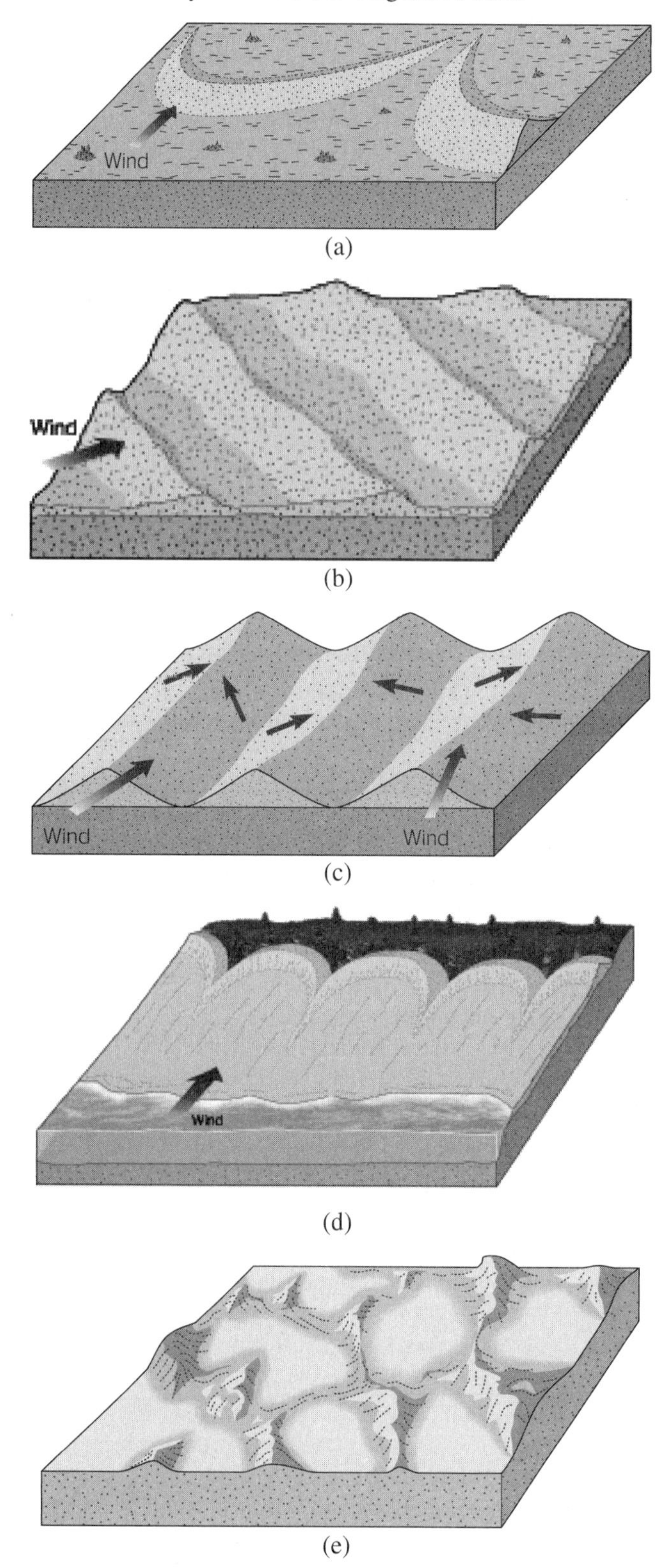

Chapter 19

Shorelines and Shoreline Processes

CHAPTER OBJECTIVES

By the end of this chapter you should be able to:

1. Discuss the cause of tides and the factors that influence them.
2. Name the parts of a wave and describe wave motion.
3. Discuss the factors that influence size of waves and those which cause a wave to break.
4. Distinguish between rip currents and nearshore currents and discuss their causes.
5. List and distinguish between the various parts of a beach.
6. Discuss the material, origin, and transportation of beach sediment.
7. Describe the differences between a "summer beach" and a "winter beach" and account for the differences.
8. Define and compare and contrast the origins of spits, baymouth bars and tombolos.
9. Describe a barrier island and list two theories for their origin..
10. List the main processes responsible for erosion along a shoreline and the landforms that result.
11. Discuss the concept of a shoreline's sediment budget and list the sources of sediment, the processes responsible for sediment removal, and the consequences of negative vs. positive budget.
12. Discuss the effects of the rising sea level on the East and Gulf Coasts and how humans are trying to deal with these effects.
13. Contrast depositional coasts with erosional coasts and emergent coasts with submergent coasts.

KEY TERMS

After reading this chapter, you should be familiar with the following terms

shoreline
tides
flood tide
ebb tide
spring tide
neap tide
wave
crest
trough
wavelength
wave height
celerity
wave period
wave base
seas (in the wave sense)
swells
fetch
breaker
nearshore zone
breaker zone
surf zone
wave refraction
longshore currents
rip currents
beach
pocket beach

backshore
berm
beach face
foreshore
longshore drift
groins
summer beach
winter beach
spit
baymouth bar
hook (or recurved spit)
tombolo
jetties
barrier island
corrosion
hydraulic action
abrasion
wave-cut platform
wave-built platform
marine terrace
headland
sea cave
sea arch
sea stacks
nearshore sediment budget
ebb tide delta
storm surge
depositional coast
erosional coast
submergent (or drowned) coast
estuary
emergent coast

CHAPTER CONCEPT QUESTIONS

1. What causes ocean tides? What unique conditions cause spring and neap tides?

2. Define wavelength, wave period and celerity. How do they relate to each other?

3. Describe the path that a "water particle" follows as waves advance across the water. How does this change with depth? How does it change as the shoreline is approached?

4. How do most waves that you see at the shore form? What factors control their size?

5. Why are waves on the ocean so much larger than those on lakes?

6. What causes a swell to change to a breaker?

7. How does the slope of the bottom offshore affect the character of a breaker? What other factors can influence breaker characteristics (particularly size)?

8. What is wave refraction and why does it happen?

9. What are longshore currents and how do they form?

10. What are rip currents and how do they form? What determines their location?

11. Explain the spatial relationships between the following terms: (a) beach, (b) foreshore, (c) backshore, (d) beach face, and (e) berms.

12. Describe the path sediment typically takes as it is transported to, and along, a beach.

13. How do seasons affect the characteristics of a beach?

14. Explain the origin of the following: (a) spit, (b) baymouth bar, and (c) tombolo.

15. What are barrier islands and where are they found? Describe two models for their origin.

16. List and define the processes that are important in shoreline erosion.

17. Explain how wave-cut and wave-built platforms develop. How do they evolve into marine terraces?

18. Explain the concept of a sediment budget for shorelines. What are the sources of sediment supplied to shorelines? What processes deplete shorelines of sediment?

19. Describe two factors that control sea level change.

20. What problem does the worldwide rise in sea level pose for society? What measures are being taken to mitigate the potential damage? What further problems have been caused by some of these measures?

21. Contrast emergent and submergent coasts.

COMPLETION QUESTIONS

1. Shorelines are defined as the areas between _______ tide and the highest level on land affected by _______ waves.

2. A rising tide is called a(n) _______ tide and a falling tide is a(n) _______ tide.

3. The highest tidal ranges occur during the _______ tides, and lower tidal ranges occur during the _______ tides.

4. When the sun and moon are aligned the result is a _______ tide, and when they are at right angles to one another the result is a _______ tide.

5. Though waves have several origins, most are generated by the _______.

6. The highest part of a wave is its _______ and the lowest is the _______. The vertical distance between these two points is the wave _______.

7. As waves pass the "particles" of water actually follow a _______ path.

8. The depth at which water "particle" motion dies out is called _______, a depth roughly equal to _______ the wavelength.

9. As the wind blows across the water beneath a storm center, it forms sharp crested choppy waves called _______.

10. As waves leave a storm generating area they begin to sort themselves out into regular _______, which have rounded crests.

11. If two or more waves of different periods happen to have their crests coincide a large _______ can develop.

12. A major control of the size to which a wave can develop is the _______, which is the distance the wind has been blowing over the water.

13. As the incoming swell approaches a beach, the wave base intersects the bottom, the waves become steeper and _______.

14. After wave base intersects the bottom, friction causes the wave velocity to _______, the wavelength to _______, and the wave height to _______.

15. Longshore currents flow _______ to the shoreline and form when waves arrive at an angle to the beach.

16. _______ currents form by the converging of two longshore currents. They flow toward the _______.

17. A _______ is a deposit of unconsolidated sediment extending landward from low tide to a major change in topography or a line of permanent vegetation.

18. The backshore is that part of a beach only covered by water during extremely high tides or _______ and consists of one or more _______, nearly horizontal platforms deposited by waves.

19. The area of a beach between high and low tide is the _______.

20. Longshore currents transport sand along a beach in a process called longshore _______.

21. Structures built to stabilize the beach by minimizing the effects of longshore currents are called _______.

22. Sandbars that are fed by longshore currents and connected to the mainland at only one end are called _______.

23. If a spit grows across an embayment, it forms a _______ bar.

24. A tombolo connects a(n) _______ to the _______.

25. Barrier islands are oriented_______ to the shoreline and are separated from the mainland by a _______.

26. As a sea cliff retreats, a gently sloping platform called a _______ platform forms.

27. Isolated columns of rock left by a receding sea cliff are called _______.

28. If the land goes up with respect to sea level, an _______ coast is formed.

29. If the land goes down with respect to sea level, a _______ coast has formed.

30. When wave-cut platforms are raised above sea level, generally by tectonic processes, they are called marine _______.

MULTIPLE CHOICE

1. In most areas high tides occur
   a. once a day.
   b. twice a day.
   c. four times a day.
   d. once every other day.

2. Spring tidal ranges occur
   a. during full and new moons.
   b. during first and last quarter moons.
   c. during April and May.
   d. none of the above

3. The distance from crest to crest is called the
   a. wave period.
   b. wave speed.
   c. wave height.
   d. none of the above

4. The sequence of wave development is
   a. sea, breaker, swell.
   b. breaker, sea, swell.
   c. sea, swell, breaker.
   d. breaker, swell, sea.

5. If the wavelength is 10 meters, the wave height is 1 meter and the wave period is 50 seconds, calculate the celerity.
   a. 10 meters
   b. 5 seconds/meter
   c. 500 meter seconds
   d. .2 meters per second

6. Waves begin to break when they are in a water depth of _____ times the wavelength.
   a. 0.5
   b. 1.25
   c. 1.0
   d. 1.5

7. The size of a wave is controlled by the
   a. wind speed.
   b. wind duration.
   c. fetch.
   d. all of the above

8. The diameters of the orbitals followed by water "particles" during wave motion ____ with depth.
   a. increase
   b. decrease
   c. are unchanged

9. Swell waves are characterized by having
   a. sharp crests and a variety of sizes.
   b. rounded crests and a consistency of size.
   c. sharp crests and a consistency of size.
   d. rounded crests and a variety of sizes.

10. When a wave refracts its crest
    a. becomes more parallel to the shoreline.
    b. becomes more perpendicular to the shoreline.
    c. begins to move away from the shoreline.
    d. disappears.

11. The position of rip currents is largely determined by
    a. the strength of the wind.
    b. the tidal range.
    c. the configuration of the sea floor.
    d. the distance to the equator.

12. Berms are generally found on the
    a. foreshore.
    b. backshore.
    c. beach face.
    d. none of the above

13. The most common mineral found in beach sands is
    a. quartz.
    b. potassium feldspar.
    c. calcite.
    d. magnetite.

14. The difference between summer and winter beaches is primarily due to
    a. seasonal variations in the strength of tides.
    b. freezing and thawing of ocean water.
    c. seasonal reversals in the direction of longshore currents.
    d. seasonal differences in the frequency and intensity of storms.

15. Spits and baymouth bars are constructed by
    a. tidal currents.
    b. longshore currents.
    c. rip currents.
    d. breakers during large storms.

16. A baymouth bar is
    a. a linear bar separated from the mainland by a lagoon.
    b. a bar that totally seals off an embayment.
    c. a bar that connects an island to the mainland.
    d. none of the above

17. As sea level rises, barrier islands
    a. migrate landward.
    b. migrate seaward.
    c. remain fixed.
    d. grow taller.

18. As a sea cliff recedes a _____ develop.
    a. tombolo
    b. a wave cut platform
    c. a sea arch
    d. barrier island

19. Which of the following correctly lists the order of formation of landforms as a headland is eroded?
    a. sea cave, sea stack, sea arch
    b. sea stack, sea cave, sea arch
    c. sea cave, sea arch, sea stack
    d. sea arch, sea cave, sea stack

20. Most sediment is transported to a beach by ________, and distributed along the shoreline by ________.
    a. streams / rip currents
    b. longshore currents / rip currents
    c. streams / longshore currents
    d. longshore currents / tides

21. Sediment is removed from shorelines by
    a. wind.
    b. longshore currents.
    c. transport by offshore directed currents.
    d. all of the above

22. Emergent shorelines are due to
    a. falling sea level due to glaciation.
    b. rising land due to local tectonic activity.
    c. rising land due to isostatic rebound.
    d. any of the above

23. Hazards along shorelines include
    a. collapse of sea cliffs.
    b. landward migration of the shore.
    c. hurricanes.
    d. all of these answers

24. In California, sediment retention dams have been built to protect communities from debris flows and floods. What effects do these dams have on the coastline?
    a. no effects
    b. They cause more barrier islands to form
    c. They cause beaches to disappear
    d. They increase the nearshore sediment budget

25. Rip currents
    a. transfer water brought in by waves back out to sea.
    b. are usually parallel to the shoreline.
    c. carry water landward in the nearshore zone.
    d. form chiefly in areas of high tidal range.

TRUE OR FALSE

___1. Shorelines are attractive to developers because they are geologically more stable than most other areas.

___2. Sea level is currently falling along most of the North American coast.

___3. Flood tides occur only during full or new moons and are about 20% higher than other high tides.

___4. Neap tides form when the Sun and the Moon are at right angles to each other with respect to Earth.

___5. Waves in lakes are generally smaller than in the ocean because the wind cannot blow as hard over land as over the sea.

___6. Wave height is measured from the sea floor to the crest of a wave.

___7. Waves "feel the bottom" when the depth is about .5 times the wavelength.

___8. As waves move toward the shore in deep water, water "particles" experience little or no net forward movement.

___9. Longshore currents form when waves approach the beach at an oblique angle to the shoreline.

___10. Rip currents flow away from the shoreline, sometimes at high velocities.

___11. The backshore area of a beach is covered during most high tides.

___12. Groins have proven nearly 100% effective in halting erosion of beaches.

____13. Typically removal of sand from a beach dominates in the winter and deposition of sand on a beach dominates in the summer.

____14. Winter beaches generally have wider berms than summer beaches.

____15. Jetties are constructed in an effort to prevent baymouth bars from blocking harbor entrances.

____16. In the United States the best-developed barrier islands are found off the coasts of Washington, Oregon, and northern California.

____17. Currently sea level, on a worldwide basis, is dropping.

____18. Due to wave refraction, wave energy is concentrated in embayments, and diffused or scattered on headlands.

____19. The presence of sea stacks is indicative of retreating cliffs in an erosion-dominated shoreline.

____20. Erosion of sea cliffs only contributes between 5 and 10 percent of the sediment supplied to shorelines.

____21. Modern coastal management techniques have proven highly successful in arresting the damage caused by rising sea level.

____22. During large storms coastal flooding causes more damage than wind.

____23. Estuaries are best developed along submergent coastlines.

____24. Marine terraces are best developed along submergent coastlines.

____25. Wind generates most surface waves.

DRAWINGS AND FIGURES

1. In the cross-section below, label a wave crest and a trough. Also indicate the wavelength and height. Indicate also the path that a water "particle" takes.

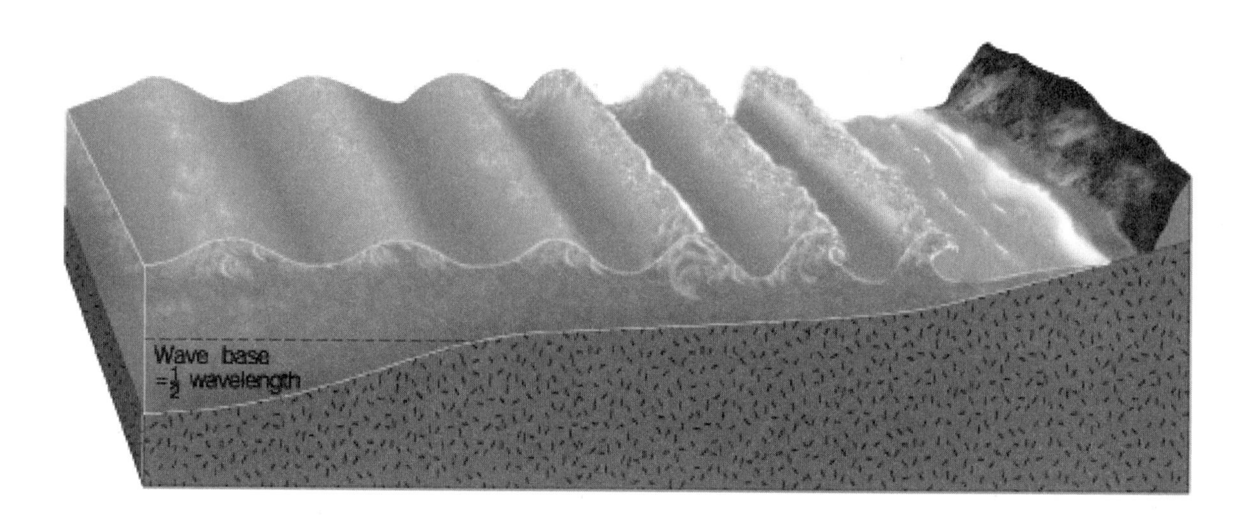

2. Identify the features indicated by an lines in the diagrams below.

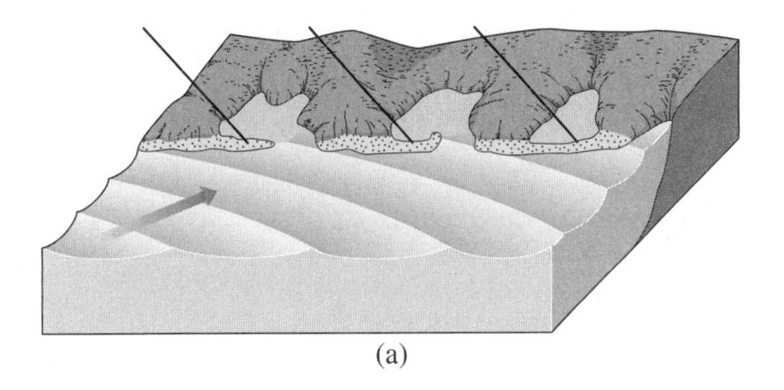

(a)

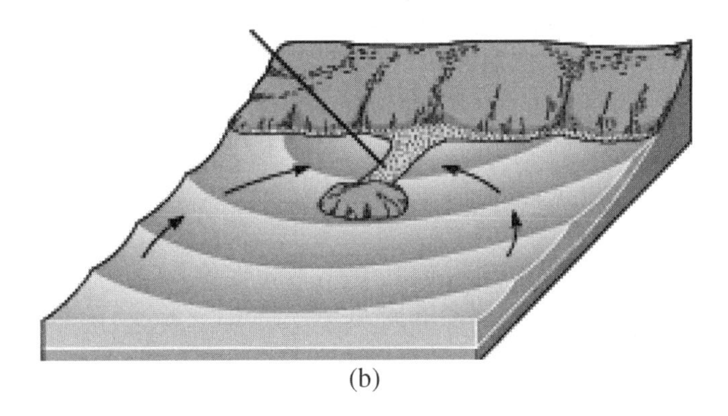

(b)

3. Identify the features indicated by an arrow in the two photographs below.

(a)

Chapter 20

Geology and Humanity

CHAPTER OBJECTIVES

By the end of this chapter you should be able to:

1. Synthesize what you have learned about how the earth works.
2. Apply that knowledge to how humanity affects the earth and vice versa.

CHAPTER CONCEPT QUESTIONS

1. What are the consequences of "the great modern expansion" according to J.R. McNeill?

2. List three ways humans alter global systems.

3. Give an example that supports Will Durant's statement: "civilization exists by geologic consent subject to change without notice."

4. What threatens stable food production today?

5. List evidence that sea level can change quickly.

COMPLETION QUESTIONS

1. People settle around volcanoes despite the danger because the volcanic ________- rich soil is very _______.

2. The Romans invented ______________ using volcanic _______ and ash.

3. People have _____________ global erosion rates by at least 10X over nature through agriculture, urbanization, _____________, and other activities.

4. ________% of Earth's land surface is used for growing food.

5. Mine tailings (waste rock) to make one gold ring amounts to 100 _______.

6. As population goes up, ________ use goes up, and human _______ on nature goes up.

7. Throughout history, geology has influenced _______ and _____ people live.

8. The earliest centers of urban- agricultural life were in places with rich ________, ________, and flooding to add nutrients.

9. The rate of soil loss is ________ greater than the rate of soil formation.

10. Some sediment and ice cores indicate CO_2 is higher now than anytime in the last _________ years.

MULTIPLE CHOICE

1. The ice age would've inhibited the development of major civilizations by
   a. making agriculture unreliable.
   b. causing sea routes to be blocked by icebergs.
   c. covering too much land with glaciers.
   d. all of these answers

2. Early settlements were subject to frequent floods because
   a. sea level rose rapidly as the glaciers retreated.
   b. people settled on agriculture-rich floodplains.
   c. desert areas with their flashflood dangers had the richest gold deposits.
   d. iron mines are invariably in valleys.

3. Some of man's earliest laws regulated
   a. concrete quarries.
   b. coal mines.
   c. irrigation and water rights.
   d. chromium ore mining.

4. When ice shelves melt,
   a. the melted ice raises sea level many meters.
   b. they refreeze immediately and kill fish.
   c. no major effects are observed.
   d. the glacier advances quickly.

5. The Romans and the Phoenicians fought for control of __________ because of its geologic resources.
   a. Burma
   b. Pompeii
   c. Spain
   d. Sumeria

6. 100,000 died in Lisbon in 1755 because of
   a. a catastrophic volcanic eruption.
   b. a tsunami.
   c. starvation due to drought.
   d. a war over geologic resources.

7. Early European settlers in North America traveled through the Appalachians using
   a. water gaps.
   b. great transform faulted valleys.
   c. the train.
   d. ancient Native American mining camps.

8. The world has nearly _______ dams.
   a. 1000
   b. 40,000
   c. 100,000
   d. 400

9. Since 1985, average annual temperature has risen at the same rate as
   a. glacier size.
   b. volcanic activity.
   c. petroleum use.
   d. all of these answers

10. The use of synthetic fertilizers has
   a. greatly increased food production.
   b. promoted algal blooms.
   c. killed aquatic life.
   d. all of these answers

TRUE OR FALSE

___1. Over-use of irrigation in deserts can cause fields to salt up.

___2. People have reduced overall erosion through paving & urbanization and through agriculture.

___3. Deforestation and soil erosion can have deadly effects.

___4. Streams carry more sediment to the sea then the amount produced by mining.

___5. Climate change is always slow and gradual on a human time-scale.

___6. Global energy use is 5X what it was a lifetime ago.

___7. The Sumerian civilization collapsed in part because of over irrigation.

___8. The most erosion occurs from agricultural and deforested land in developing countries.

___9. Humans add methane to the atmosphere by cutting down forests.

___10. If ice shelves melt away, the Antarctic glaciers would move and melt faster.

DRAWINGS AND FIGURES

1. What would be some of the consequences of building a dam in the headwaters of this river?

2. Draw and explain why the loss of ice shelves would increase global sea level.

Appendix

Answers to Questions

Answers to Chapter Concept Questions: Because a primary goal of a study guide is to encourage study we are not providing detailed answers to these questions. To do so would encourage memorization of material that should be studied and understood and would, therefore, be counterproductive A student should study the material in the text and formulate his/her own answers. We have, however, provided page numbers and figure and table references so that the student can readily find information necessary to answer the question. In some cases some additional hints are provided.

Answers to Tables: Filling in the tables in these questions is an excellent way to study some of the details of the chapters that many instructors expect students to know. Therefore, we are not providing detailed answers to these tables. Once again we are providing page numbers and table references where the pertinent information can be found.

Answers to Completion Questions: The answers provided are the preferred answers for the blanks in the questions. Answers for questions with multiple blanks are separated by a "/". Answers to these questions are given in the same order in which the blanks occur in the questions, but in some cases different order might be acceptable such as when the blanks are simply a list. The student should be able to tell when that situation occurs. In some cases more than one answer is possible and we have included those alternative answers in parentheses.

Answers to Multiple Choice and True or False Questions: The correct answers are provided.

Answers to Drawings and Figures: Because drawing talents vary so widely we are not providing detailed answers to these questions. To do so would take up too much space and might discourage those with limited drawing skills. In some cases you are simply asked to identify or label the features in a photograph or drawing. Depending on the complexity of the question either an answer or a page number or a figure reference is provided.

Chapter 1

Chapter Concept Questions

1. p. 4, Table 1.1, Fig. 1.1
2. p. 6-7, Table 1.2
3. p. 7-8
4. p. 8
5. p. 8-10
6. p. 9-10, Fig. 1.5
7. p. 10-11
8. p. 10-11
9. p. 11-12
10. p. 12, 14-15
11. p. 12-13, 16
12. p. 16-17
13. p. 17-18, Fig. 1.11, 1.12
14. p. 16, Fig. 1.10
15. p. 13, 19-21
16. Table 1.3, p. 18
17. p. 19-21, Fig. 1.12
18. p. 22-23

Complete the Following Tables See pages 12 & 16 for the pertinent data for both tables.

Completion Questions

1. minerals / rocks
2. layered rocks
3. deformation
4. fossils
5. carbon dioxide
6. 15 billion
7. hydrogen / helium
8. 4.6
9. solar nebula
10. sun / planetesimals (planets)
11. hypothesis / theory
12. Plate Tectonic
13. convection / mantle
14. away from
15. Sedimentary
16. Metamorphic
17. Uniformitarianism
18. volcanoes/ earthquakes
19. peridotite/ mantle
20. dynamic/ complex
21. overpopulation/ six
22. Si/ Fe/ Mg/ O or Al
23. geologic resources or materials
24. densities/ pressures
25. igneous/ converge

Multiple Choice

1. d
2. b
3. d
4. c
5. b
6. d
7. d
8. d
9. c
10. d
11. b
12. d
13. d
14. b
15. a
16. b
17. c
18. c
19. d
20. d
21. a
22. a
23. d
24. b
25. c

True or False

1. T
2. F
3. F
4. T
5. T
6. F
7. F
8. T
9. T
10. T
11. F
12. F
13. T
14. F
15. F
16. F
17. F
18. F
19. T
20. T
21. T
22. F
23. T
24. F
25. T

Drawings and Figures

1. Fig. 1.13
2. Fig. 1.9

Chapter 2

Chapter Concept Questions

1. A lengthy answer using information from pages 36-40
2. p. 40
3. p. 40-41
4. Continental Drift is summarized p. 36-40; Seafloor Spreading is summarized p. 41-42.
5. p. 43, Fig. 2.10
6. p. 44, Fig. 2.11
7. p. 46-47
8. p. 46-47
9. p. 47-49, Table 2.1, Fig. 2.15
10. p. 47-49, Fig. 2.15
11. p. 52
12. p. 52-54, Figs. 2.18
13. p. 54
14. p. 54, Fig. 2.19
15. p. 58
16. p. 59-60
17. p. 60-61

Completion Questions

1. Glossopteris
2. Pangaea / Laurasia / Gondwana
3. Alfred Wegner
4. 2000 meters
5. South Africa
6. Sea-floor spreading
7. Magnetic anomalies / mid-ocean ridge
8. 180 million / 3.96 billion
9. oceans/ climate
10. forms / destroyed (subducted)
11. lithosphere / asthenosphere
12. older / greater
13. basalt
14. rift valley
15. Benioff
16. intermediate
17. island arc
18. back-arc basin
19. subduction
20. melange
21. ophiolites
22. San Andreas
23. spots / magma
24. Hawaii
25. convection

Multiple Choice

1. d	6. c	11. c	16. a	21. d
2. c	7. a	12. d	17. c	22. d
3. b	8. d	13. d	18. a	23. c
4. a	9. b	14. c	19. b	24. c
5. d	10. b	15. b	20. d	25. d

True or False

1. T	6. F	11. F	16. F	21. T
2. F	7. T	12. T	17. T	22. F
3. F	8. F	13. T	18. F	23. F
4. T	9. F	14. T	19. F	24. T
5. T	10. T	15. F	20. F	25. T

Drawings and Figures

1. Fig. 2.14
2. Fig. 2.15
3. Figs. 2.18

Chapter 3

Chapter Concept Questions

1. p. 72-73, 77
2. p. 73
3. p. 74-75
4. p. 75
5. p. 75-77
6. Glass, water and bone are not (they are not crystalline, not solid, organic respectively). Ice and quartz are minerals because they are crystalline solids.
7. p. 77
8. p. 78
9. Fig. 3.10, p.79
10. p. 80
11. Silicon is an element, silica is a combination of silicon and oxygen, silicate is a mineral family. See p. 80.
12. p. 80-81, Fig. 3.13
13. p. 85, 88-90
14. p. 89
15. p. 91
16. p. 90
17. p. 92
18. p. 92-93

Completion Questions

1. naturally / crystalline
2. Rocks
3. matter
4. elements / atoms
5. nucleus / protons / neutrons
6. atomic
7. protons / neutrons
8. protons / neutrons
9. bonding
10. compound
11. ion
12. ionic / covalent
13. metallic
14. native elements
15. ions
16. oxygen / silicon / silicates
17. dark
18. iron / magnesium
19. carbonate
20. physical
21. color (luster)
22. cleavage / fracture
23. hardness
24. water
25. rock / silicate

Multiple Choice

1. d
2. c
3. b
4. c
5. a
6. c
7. c
8. b
9. c
10. a
11. d
12. a
13. c
14. b
15. c
16. c
17. b
18. d
19. b
20. c
21. c
22. c
23. d
24. c
25. d

True or False

1. F
2. T
3. T
4. T
5. F
6. T
7. T
8. F
9. T
10. F
11. F
12. T
13. F
14. T
15. T
16. T
17. F
18. T
19. F
20. T
21. T
22. T
23. F
24. F
25. F

Drawing Question

1. Fig. 3.13
2. Fig. 3.17

Chapter 4

Chapter Concept Questions

1. p. 113
2. p. 108-109
3. p. 104
4. p. 104
5. p. 105-106, 109-111
6. p. 113-120
7. p. 120
8. p. 120-122
9. p. 115
10. p. 123-124
11. p. 104-113
12. p. 105
13. p. 105

Complete the Following Tables Table 4.2, pages 115-120, and Fig. 4.18 all contain useful information.

Completion Questions

1. extrusive / intrusive
2. calcium / sodium
3. pressure ridges
4. 25
5. low / high
6. aphanitic
7. phaneritic
8. porphyritic
9. vesicular
10. pyroclastic or fragmental
11. texture / composition
12. 65
13. rhyolite
14. diorite
15. Basalt
16. olivine
17. pegmatite
18. granite
19. tuff
20. Obsidian / pumice
21. pluton (intrusion)
22. dike / sill
23. laccoliths
24. Magmatic stoping
25. granitic (felsic) / dioritic (intermediate)

Multiple Choice

1.	d	6.	b	11.	d	16.	d	21.	a
2.	d	7.	c	12.	a	17.	a	22.	c
3.	a	8.	c	13.	a	18.	d	23.	c
4.	d	9.	b	14.	b	19.	b	24.	a
5.	d	10.	c	15.	a	20.	b	25.	b

True or False

1.	F	6.	F	11.	F	16.	T	21.	T
2.	T	7.	T	12.	F	17.	T	22.	F
3.	F	8.	F	13.	T	18.	T	23.	T
4.	F	9.	F	14.	F	19.	T	24.	T
5.	F	10.	T	15.	F	20.	F	25.	T

Drawing and Figures

1. The give-away answer is Fig. 2 on p. 129. See also Fig. 4.24, p. 123
2. (a) rhyolite is aphanitic, granite is phaneritic; (b) basalt is aphanitic, gabbro is phaneritic; (c) andesite is aphanitic, diorite is phaneritic. See Fig. 4.16, p. 115
3. See Fig. 4.24, p. 123.

Chapter 5

Chapter Concept Questions

1. p. 136
2. p. 136-137
3. p. 137-138
4. (a) through (c) are discussed on p. 138. Pillow lavas (d) are on p. 139. Calderas (e) are described on p. 142. Lava domes (f) are on p. 144-145. Welded tuff (g) is on p. 150. Basalt plateaus (h) are on p. 149.
5. p. 139
6. p. 141,143
7. p. 141-142,144
8. p. 144, 146
9. p. 153-155
10. p. 158
11. p. 157-159
12. p. 151-152, 156
13. p. 156

Completion Questions

1. dormant / extinct (inactive)
2. water vapor
3. Io/ Earth/ possibly Titan
4. columnar joints
5. pillow lava
6. ash
7. volcanoes / fissures
8. crater
9. caldera
10. shield / basaltic (mafic)
11. Kilauea
12. cinder
13. composite
14. lahars
15. shield
16. composite
17. basalt plateaus
18. pyroclastic sheet deposit
19. 8 /7
20. harmonic tremor
21. mid-ocean ridges
22. subduction zones
23. composite
24. columnar joints
25. low/ expand/ escape

Multiple Choice

1. a	6. c	11. c	16. b	21. c
2. b	7. c	12. c	17. d	22. a
3. b	8. a	13. a	18. d	23. b
4. d	9. c	14. a	19. b	24. d
5. a	10. b	15. b	20. a	25. d

True or False

1. T	6. T	11. T	16. F	21. T
2. F	7. T	12. F	17. F	22. F
3. F	8. F	13. T	18. F	23. F
4. F	9. F	14. T	19. F	24. F
5. T	10. T	15. T	20. F	25. F

Drawings and Figures

1. Fig. 5.9
2. Fig. 5.10, 5.11, 5.13
3. Fig. 5.23 a. 2, b. 3, c. 6

Chapter 6

Chapter Concept Questions

1. p. 171-172
2. p. 172
3. p. 173
4. p. 173
5. p. 174
6. p. 175, 178, Fig. 6.7
7. p. 178
8. p. 178-179
9. p. 180
10. p. 180, Table 6.1
11. p. 183-184, Fig. 6.16
12. p. 183-184
13. p. 183-186
14. p. 187-190
15. p. 191

Completion Questions

1. weathering
2. erosion
3. differential
4. chemical / mechanical
5. frost
6. pressure / eroded
7. water
8. solution / oxidized / hydrolysis
9. clay
10. Small (Clay)
11. spherical (rounded)
12. faster
13. soils
14. organic
15. transported
16. A / B
17. accumulation / leaching
18. climate
19. pedalfer
20. caliche / pedocals
21. laterite / bauxite
22. deeper (thicker)
23. expansive
24. wind / water
25. residual

Multiple Choice

1.	c	6.	c	11.	c	16.	c	21.	a
2.	b	7.	d	12.	d	17.	d	22.	c
3.	c	8.	b	13.	c	18.	b	23.	a
4.	c	9.	c	14.	b	19.	d	24.	d
5.	d	10.	b	15.	b	20.	b	25.	c

True or False

1.	F	6.	T	11.	T	16.	F	21.	T
2.	T	7.	F	12.	F	17.	F	22.	T
3.	T	8.	F	13.	T	18.	F	23.	F
4.	F	9.	T	14.	F	19.	T	24.	F
5.	T	10.	T	15.	T	20.	F	25.	T

Drawing and Figures

1. (a) talus, explained p. 172, Fig. 6.3c; (B) exfoliation dome; explained p. 173, Fig. 6.4.
2. Fig. 6.14

Chapter 7

Chapter Concept Questions

1. p. 201-202
2. p. 203-204
3. p. 205
4. p. 201, 204-206
5. p. 206
6. p. 206-208
7. p. 208
8. p. 208-210
9. p. 210-211, especially Figs. 7.12
10. p. 211-212
11. p. 213, Figures 7.14 through 7.18
12. p. 216-217
13. p. 214-215, Geology in focus on p. 226
14. p. 215, 218
15. p. 219-221
16. p. 219-220

Complete the Following Tables Table 7.1, Fig. 7.4, and pages 204-209

Completion Questions

1. weathering (disintegration)
2. detrital
3. water
4. abrasion
5. sorting
6. depositional environment
7. lithification
8. compaction / cementation
9. hematite
10. clasts / clastic
11. gravel / 2
12. rounding
13. quartz
14. claystone (mudrock, shale)
15. fissility
16. coquina
17. Ooids
18. evaporation
19. quartz (silica)
20. Coal / oxygen
21. lignite / bituminous / anthracite
22. transgression / regression
23. structures
24. deposition
25. Cross-bedding
26. mold / cast
27. Coal / oil / natural gas (uranium)
28. source / reservoir
29. Persian Gulf
30. kerogen
31. Precambrian

Multiple Choice

1. c
2. b
3. a
4. b
5. a
6. c
7. d
8. d
9. b
10. a
11. c
12. d
13. a
14. a
15. c
16. b
17. a
18. b
19. d
20. d
21. c
22. c
23. d
24. a
25. a

True or False

1. T
2. F
3. F
4. T
5. T
6. T
7. T
8. T
9. T
10. F
11. F
12. F
13. T
14. F
15. T
16. T
17. F
18. T
19. T
20. F
21. T
22. T
23. F
24. T
25. T

Drawings and Figures

1. cross-bedding, right to left, Fig. 7.14
2. Fig. 7.3

3. Fig. 7.12a is transgression, Fig. 7.12b is regression
4. (See Fig. 7.17) (a) ripple marks, (b) mud cracks

Chapter 8

Chapter Concept Questions

1. p. 233-234
2. p. 235-236, Fig. 8.5
3. p. 237-238, Fig. 8.7, 8.10
4. p. 234-235
5. p. 235-238
6. p. 235-237
7. p. 237-238
8. p. 239-240
9. p. 248, Fig. 8.18
10. p. 241-242, especially Fig. 8.10
11. p. 248-249
12. p. 253-254, 258, Table 8.1

Complete the Following Tables Table 8.2 and pages 241-247

Completion Questions

1. mineral / texture
2. shields
3. pressure / temperature / fluids
4. magma / geothermal
5. Lithostatic
6. differential
7. increases
8. pore spaces / magma / dehydration
9. contact / regional / dynamic
10. Contact
11. aureoles
12. Dynamic
13. regional
14. differential
15. Slate
16. Gneiss
17. Migmatites
18. calcite / limestone
19. quartzite
20. contact
21. anthracite
22. facies
23. isograds
24. dynamic/ faults
25. index/ high

Multiple Choice

1. d
2. b
3. c
4. c
5. a
6. b
7. b
8. b
9. a
10. c
11. d
12. c
13. b
14. c
15. a
16. b
17. a
18. a
19. d
20. b
21. b
22. d
23. a
24. c
25. b

True or False

1. F
2. F
3. T
4. T
5. F
6. F
7. F
8. T
9. T
10. T
11. F
12. F
13. F
14. F
15. F
16. T
17. T
18. T
19. T
20. F
21. T
22. F
23. F
24. F
25. T

Drawings and Figures

1. See Fig. 8.3 and 8.21 in text.
2. (a) blueschist; (b) sanidinite, amphibolite, greenschist, zeolite; (c) granulite, amphibolite, greenschist, pumpellyite or zeolite

---

Chapter 9

Chapter Concept Questions

1. p. 262
2. p. 264
3. p. 264-265
4. p. 265-266
5. p. 267-268, Fig. 9.3, 9.4, 9.5
6. p. 266, Fig. 9.6
7. p. 266-267, especially Figs. 9.4, 9.8, 9.9, 9.10
8. p. 278, 280
9. p. 280
10. p. 281, Fig. 9.18
11. p. 282-284
12. p. 284
13. p. 285
14. p. 285-286, especially Fig. 9.24
15. p. 286-287, especially Fig. 9.25

Completion Questions

1. Relative
2. Absolute
3. 4.6 billion
4. Uniformitarianism
5. bottom
6. sediment
7. younger
8. older
9. unconformities
10. angular
11. sedimentary / igneous / metamorphic
12. fossils
13. key beds
14. wide (large) / short (small, limited)
15. concurrent range zones
16. radioactive (parent) / daughter
17. protons / two
18. neutron / proton / electron
19. half life
20. mass spectrometer
21. 70,000
22. nitrogen-14 / beta
23. rings
24. relative/ absolute
25. sedimentary/ layers/ fossils

Multiple Choice

1.	c	6.	a	11.	c	16.	a	21.	c
2.	d	7.	b	12.	d	17.	a	22.	c
3.	a	8.	b	13.	d	18.	d	23.	d
4.	c	9.	d	14.	a	19.	b	24.	a
5.	b	10.	a	15.	b	20.	b	25.	b

True or False

1.	F	6.	T	11.	T	16.	T	21.	T
2.	T	7.	T	12.	F	17.	T	22.	T
3.	T	8.	F	13.	F	18.	F	23.	T
4.	F	9.	T	14.	F	19.	F	24.	F
5.	T	10.	T	15.	F	20.	T	25.	T

Drawings and Figures

1. (a) through (c) review Steno's principles, p. 265-266, especially Fig. 9.2.
2. Figs. 9.8, 9.9, 9.10
3. Figs. 9.11, 9.12 and the discussion which accompanies these on pages 271-273
4. Between 120 and 250 million years old

Chapter 10

Chapter Concept Questions

1. p. 301-302, Fig. 10.1
2. p. 303, Fig. 10.3
3. p. 303-304, Fig. 10.5
4. p. 307-308, Fig. 10.8 & 10.9
5. p. 307-308, especially Fig. 10.8 & 10.9
6. p. 309-310, especially Fig. 10.10
7. p. 309-310, especially Fig. 10.11
8. p. 311, Table 10.2
9. p. 311, 313, especially Fig. 10.13
10. p. 311, especially Fig. 10.12
11. You will have to do some thinking here. Take notes as you read p. 316-322 and develop your own ideas. The Geology in focus on p. 331 will provide some insight.
12. p. 317-322
13. p. 317-318
14. p. 323-325
15. p. 324-325, especially Fig. 10.23
16. p. 326-327

Completion Questions

1. energy
2. elastic rebound
3. seismology
4. seismograph / seismogram
5. focus (hypocenter) / epicenter
6. 70 / 300
7. Benioff
8. circum-Pacific
9. Mediterranean - Asiatic
10. 900,000
11. body / surface
12. primary (P) / secondary (S)
13. primary (P) / parallel
14. Secondary (S) / perpendicular
15. Love (L) / Rayleigh (R)
16. primary (P) / secondary (S)
17. intensity / Mercalli
18. magnitude / Richter
19. 100
20. liquefaction
21. tsunamis
22. seismic risk
23. precursors
24. shallow/ most
25. weaker/ wider or greater

Multiple Choice

1.	c	6.	d	11.	c	16.	b	21.	d
2.	a	7.	a	12.	a	17.	d	22.	c
3.	a	8.	d	13.	d	18.	c	23.	a
4.	c	9.	b	14.	d	19.	d	24.	c
5.	a	10.	c	15.	b	20.	c	25.	d

True or False

1.	F	6.	F	11.	T	16.	F	21.	T
2.	T	7.	T	12.	F	17.	F	22.	T
3.	F	8.	F	13.	F	18.	T	23.	F
4.	T	9.	F	14.	F	19.	F	24.	F
5.	F	10.	T	15.	F	20.	F	25.	T

Drawings and Figures

1. Fig. 10.5 and relevant text
2. approx. 3500 km
3. about 3.6

Chapter 11

Chapter Concept Questions

1. p. 338-339
2. p. 338-339
3. p. 343-344
4. p. 344
5. p. 340-341
6. p. 341-342
7. p. 341-342
8. p. 340
9. p. 344-345
10. p. 345, 348
11. p. 348-349, especially Fig. 11.13
12. p. 349-351, especially Fig. 11.16
13. p. 352
14. p. 352-353
15. p. 352-353
16. p. 349
17. p. 353

Complete the Following Tables Table on p. 336 and information from the text p. 340-344

Completion Questions

1. 5.52 g/cm^3 / 2.5 / 3.0 g/cm^3
2. seismic waves
3. density / elasticity
4. velocity / direction
5. density / elasticity /reflected
6. seismic tomography
7. seismic tomography
8. shadow
9. 103 /143 / focus
10. 103
11. sulfur
12. mantle
13. Moho (Mohorvicic discontinuity)
14. mineral
15. kimberlite pipes
16. oceanic / continental
17. basaltic (mafic) / granitic (felsic)
18. 25 °C
19. increases / decreases
20. gravimeter
21. anomaly
22. isostacy / mantle
23. sink / rise
24. Curie
25. outer core
26. polarity
27. inclination / declination
28. anomalies
29. Paleomagnetism
30. normal / reverse

Multiple Choice

1. a
2. b
3. c
4. a
5. d
6. b
7. d
8. c
9. a
10. c
11. d
12. b
13. c
14. d
15. c
16. d
17. c
18. c
19. a
20. a
21. d
22. c
23. a
24. a
25. d

True or False

1. T
2. T
3. T
4. T
5. F
6. F
7. F
8. T
9. F
10. F
11. T
12. T
13. T
14. T
15. T
16. T
17. T
18. F
19. F
20. T
21. F
22. F
23. T
24. T
25. T

Drawings and Figures

1. Fig. 11.1
2. Fig. 11.5

3. Fig. 11.9
4. Fig. 11.13; Diagram b) doesn't really have a magnetic anomaly. The fault in Figure 10.21 involves basalts, whereas the one in this question involves continental crust. If the faulted diagram were like that in 10.21, a positive gravity and magnetic anomaly would exist over the basalt side of the fault.

Chapter 12

Chapter Concept Questions

1. p. 362-363
2. p. 364
3. p. 364-366
4. p. 366-367
5. p. 367-369
6. p. 369, especially Fig. 12.13
7. p. 369-370
8. p. 370-371
9. p. 376
10. p. 377-378, Fig. 12.20
11. p. 380-381

Completion Questions

1. *Glomar Challenger*
2. seismic
3. ophiolite
4. shelf / slope / rise
5. canyons
6. turbidity / submarine fans / rise
7. abyssal
8. Active / trenches
9. shelves / rise
10. passive / active
11. abyssal plains
12. trenches
13. 65,000
14. rift valleys
15. black smokers (hydrothermal vents)
16. fractures
17. seamounts / guyots
18. aseismic
19. pelagic
20. Ooze / calcareous
21. Reefs
22. fringing / barrier
23. Exclusive Economic
24. tensional/ mafic
25. siliceous/ skeletons (remains)

Multiple Choice

1. d	6. d	11. c	16. a	21. d
2. c	7. a	12. b	17. c	22. d
3. d	8. c	13. a	18. c	23. e
4. c	9. d	14. b	19. c	24. c
5. d	10. b	15. d	20. c	25. b

True or False

1. F	6. F	11. T	16. T	21. F
2. F	7. F	12. F	17. F	22. T
3. T	8. F	13. T	18. T	23. F
4. T	9. F	14. T	19. T	24. F
5. T	10. T	15. T	20. T	25. T

Drawings and Figures

1. Fig. 12.6, p. 364
2. Fig. 12.17, p. 376.

Chapter 13

Chapter Concept Questions

1. p. 391-392
2. p. 392-393
3. p. 393
4. p. 394-396, Figs. 13.6, 13.7
5. p. 396, Figs. 13.12, 13.11
6. p. 396, Fig. 13.11
7. p. 397, Fig 13.13
8. p. 398-400
9. p. 401-403, especially Figs. 13.15, 13.16
10. p. 404, Fig. 13.17
11. p. 403
12. p. 403, Fig. 13.17 (e)-(g)
13. p. 404-409
14. p. 407-409
15. p. 409, 413
16. p. 413-414

Completion Questions

1. force
2. compressional / tensional / shear
3. brittle / folds
4. strike / dip
5. anticline / syncline
6. axial plane / limb
7. asymmetric / recumbent
8. horizontal / inclined
9. dome / basin
10. younger
11. fault / joint
12. down
13. hanging wall / foot wall
14. reverse / 45
15. Shear
16. vertical / horizontal
17. anticline or dome
18. horsts / grabens
19. orogeny
20. Circum-Pacific / Alpine-Himalayan
21. accretionary wedge
22. ocean / continent
23. continent / continent
24. microplates
25. shields / platforms

Multiple Choice

1. b
2. d
3. b
4. b
5. a
6. a
7. b
8. a
9. c
10. d
11. d
12. c
13. a
14. b
15. d
16. a
17. d
18. c
19. a
20. d
21. b
22. d
23. c
24. d
25. c

True or False

1. T
2. T
3. F
4. F
5. T
6. T
7. T
8. T
9. T
10. F
11. F
12. F
13. T
14. T
15. T
16. F
17. T
18. T
19. F
20. T
21. T
22. F
23. F
24. F
25. T

Drawings and Figures

1. Fig. 13.8
2. (a) symmetric or upright anticline; (b) symmetric or upright syncline; (c) recumbent anticline; (d) overturned anticline; (e) overturned syncline
3. (a) reverse fault; (b) normal fault
4. right-lateral

Chapter 14

Chapter Concept Questions

1. p. 426-427
2. p. 427
3. p. 427
4. p. 427-429, 432
5. p. 433; examples from p. 433-443
6. p. 433-443, Table 14.2
7. 433, Fig. 14.7
8. p. 434, 436-437, Table 14.2
9. p. 437-438, Table 14.2
10. p. 439, Table 14.2
11. p. 443
12. p. 443, especially Fig. 14.23
13. p. 446, Fig. 14.22
14. p. 447-450

Completion Questions

1. shear / gravity
2. repose / 25 / 40
3. less
4. friction
5. parallel
6. earthquakes / water
7. rapid
8. Talus
9. slump
10. rock slide
11. flow
12. dynamic/ change
13. Debris
14. slowly
15. Quick
16. permafrost / solifluction
17. creep
18. complex
19. slope stability
20. benching
21. thick/ steep/ clay
22. steepens or undermines
23. shear strength (friction & cohesion)/ driving force (or mass or weight)
24. cloudbursts/ vegetation
25. fail/ unweathered solid

Multiple Choice

1. d	6. c	11. a	16. b	21. d
2. d	7. a	12. d	17. c	22. c
3. c	8. c	13. c	18. c	23. c
4. d	9. a	14. b	19. b	24. a
5. a	10. c	15. d	20. c	25. b

True or False

1. F	6. F	11. F	16. F	21. T
2. T	7. T	12. T	17. T	22. T
3. F	8. F	13. T	18. T	23. F
4. T	9. F	14. T	19. F	24. T
5. T	10. F	15. F	20. F	25. T

Drawings and Figures

1. Figs. 14.12, 14.17, 14.22, 14.10
2. Fig. 14.4c, p. 429

Chapter 15

Chapter Concept Questions

1. p. 459-460, Fig. 15.3
2. p. 460-461, Fig. 15.4
3. p. 461-462
4. p. 462, Fig. 15.5
5. p. 463-464, Fig. 15.6
6. p. 463-464
7. p. 463-464
8. p. 464
9. p. 464-466
10. p. 465-466
11. p. 467
12. p. 467-468, Figs. 15.11, 15.13
13. p. 467-468, Fig. 15.11
14. p. 468-469
15. p. 469-471
16. p. 471, 473
17. p. 474-476
18. p. 476-477
19. p. 478-479, Fig. 15.23
20. p. 479-481
21. p. 480-481, 484, Figs. 15.25 & 15.26
22. p. 484-485
23. p. 484-485, Fig. 15.28
24. p. 485-488
25. p. 488-489

Completion Questions

1. 97.2 / 2.15
2. 15 / 20
3. Turbulent
4. infiltration capacity
5. sheet / channel
6. gradient
7. time
8. incised meanders
9. friction
10. Discharge
11. Superposed
12. hydraulic
13. abrasion
14. bed / suspension / dissolved
15. Competency / capacity
16. channels / sediment
17. oxbow
18. cut bank / point bar
19. floodplain
20. base level
21. standing (still) / decreases
22. progrades
23. distributaries
24. stream / bird
25. alluvial fan
26. hydrograph / time
27. drainage / divides
28. dendritic
29. radial
30. joints (fractures, faults)
31. base level
32. sea level
33. Graded
34. downcutting / headward erosion

Multiple Choice

1. d
2. b
3. c
4. c
5. d
6. c
7. a
8. c
9. c
10. a
11. d
12. a
13. a
14. b
15. c
16. d
17. c
18. a
19. d
20. d
21. a
22. a
23. b
24. a
25. c

True or False

1. F
2. F
3. T
4. T
5. T
6. T
7. F
8. F
9. F
10. T
11. F
12. T
13. T
14. F
15. F
16. F
17. T
18. T
19. T
20. T
21. F
22. T
23. F
24. T
25. T
26. T

Drawings and Figures

1. 15.3
2. (a) at least 30 cm/sec, 120 cm/sec, and 40 cm/ sec respectively. Note that the Y axis is logarithmic. These answers are, therefore, approximate.
   (b) gravel at over 100 cm/s, sand at 10-15 cm/s, coarse silt at 0.5 cm/s, clay stays in suspension unless water stops moving.
3. Fig. 15.11
4. (a) a 2 year period, (b) a 4 year period, (c) a 30 year period
5. Fig. 15.23
6. The feature is an alluvial fan (Fig. 15.17a).

Chapter 16

Chapter Concept Questions

1. This is best considered and answered after reading the whole chapter but p. 498 makes some good points in the Introduction.
2. p. 498-499
3. p. 499-500
4. p. 499, especially Fig. 16.1
5. p. 500-501, especially Figs. 16.2, 16.3
6. p. 502-503
7. p. 502-503, Fig. 16.4
8. p. 503, especially Fig. 16.5
9. p. 503-504, especially Fig. 16.6
10. p. 505-509, Fig. 16.9
11. p. 505-506
12. p. 506-507, 510, especially Fig. 16.11
13. p. 511
14. p. 511-512, Fig. 16.14
15. p. 512-513, 516
16. p. 516-517
17. p. 518-521
18. p. 521, 523

Completion Questions

1. 22
2. Porosity / permeability
3. interconnected
4. aquifer / aquiclude
5. water table
6. capillary fringe
7. gravity
8. high / low
9. water table
10. perched aquifer
11. perched
12. saturation
13. depression
14. aquicludes / tilted / precipitation
15. artesian pressure
16. carbon dioxide
17. limestone / carbonic
18. sinkhole
19. saturation
20. stalactites / stalagmites / column
21. 30
22. saltwater incursion
23. subsidence
24. contamination
25. 37
26. magma
27. sinter
28. steam (gas)
29. geothermal

Multiple Choice

1. c
2. d
3. b
4. a
5. a
6. c
7. a
8. c
9. b
10. b
11. b
12. a
13. a
14. c
15. b
16. d
17. a
18. d
19. b
20. b
21. c
22. a
23. b
24. c
25. a

True or False

1. F
2. T
3. F
4. T
5. T
6. T
7. T
8. F
9. F
10. F
11. F
12. F
13. F
14. T
15. T
16. F
17. F
18. T
19. T
20. F
21. F
22. F
23. T
24. F
25. T

Drawings and Figures

1. Fig. 16.3
2. Figs. 16.11 & 16.12

Chapter 17

Chapter Concept Questions

1. p. 533-534
2. p. 536, Fig. 17.4
3. p. 536
4. p. 533-534
5. p. 538-539
6. p. 536, especially Fig. 17.5
7. p. 536-537, Fig. 17.8
8. p. 540
9. p. 541-543
10. p. 543
11. p. 543-546
12. p. 546
13. p. 547, 550
14. p. 548-549
15. p. 550
16. p. 543, 546, 550, 555-556
17. p. 554-555
18. p. 556-557
19. p. 557-558
20. p. 558-559, Fig. 17.27
21. p. 558-559

Completion Questions

1. snow
2. one tenth (10%)/ Australia
3. Antarctica / Greenland
4. 2.15
5. calving
6. sublimation
7. compacted / firn
8. Plastic
9. friction / basal slip
10. Valley
11. 3000
12. accumulation / wastage / firn
13. snow / melting / sublimation
14. retreat / advance / stagnant
15. accumulation
16. crevasses
17. surge
18. Plucking / roche moutonée
19. polish / striations
20. flour
21. U
22. fiords
23. cirque / arête
24. horn
25. continental / deranged
26. drift
27. erratics
28. till / stratified
29. end / terminal
30. recessional
31. medial
32. drumlins
33. outwash / valley
34. kettle
35. eskers
36. kames
37. dropstones / icebergs
38. Milankovitch
39. Four / 20
40. rebound
41. pluvial
42. glacier / moraine
43. 18
44. 130
45. rise / 70
46. 300

Multiple Choice

1. a
2. a
3. b
4. b
5. c
6. c
7. a
8. a
9. d
10. c
11. b
12. b
13. a
14. b
15. c
16. c
17. a
18. a
19. b
20. b
21. d
22. d
23. c
24. d
25. c
26. b

True or False

1. F
2. F
3. T
4. T
5. F
6. F
7. T
8. T
9. F
10. T
11. F
12. F
13. T
14. F
15. F
16. F
17. F
18. F
19. T
20. T
21. F
22. T
23. F
24. F
25. F
26. T

Drawings and Figures

1. Fig. 17.13
2. Fig. 17.18

Chapter 18

Chapter Concept Questions

1. p. 569-570
2. p. 569-570
3. p. 570-571
4. p. 571-573
5. p. 574-575, Fig. 18.9
6. p. 574
7. p. 575-578, especially the Figures
8. p. 578
9. p. 580-581, Fig. 18.17
10. p. 581, 584
11. p. 584, Fig. 18.17, 18.19
12. p. 584-586
13. p. 586
14. p. 586
15. p. 587
16. p. 587-590

Completion Questions

1. suspension / saltation
2. ventifacts
3. deflation
4. deflation hollow / desert pavement
5. windward (upwind) / leeward (downwind)
6. 30 / 34 / repose
7. barchan / parabolic
8. transverse / longitudinal
9. Loess
10. 10 / 30
11. Coriolis / right / left
12. 20 / 30
13. rainshadow / east
14. Mechanical / temperature / frost
15. iron / manganese
16. internal / playa
17. fan / mountain valley
18. bajada
19. pediments
20. inselbergs
21. mesas / butte
22. thin/ patches
23. larger/ buried
24. Africa/ Middle East
25. shape/ direction

Multiple Choice

1. a	6. d	11. b	16. b	21. c
2. b	7. a	12. a	17. b	22. b
3. d	8. d	13. d	18. d	23. d
4. a	9. d	14. d	19. b	24. d
5. b	10. a	15. b	20. c	25. a

True or False

1. T	6. F	11. F	16. F	21. F
2. F	7. F	12. F	17. T	22. T
3. F	8. F	13. T	18. T	23. T
4. T	9. T	14. T	19. T	24. T
5. T	10. F	15. F	20. T	25. F

Drawings and Figures

1. Fig. 18.2
2. Figs. 18.11, 18.13, 18.12, 18.14, 18.15

Chapter 19

Chapter Concept Questions

1. p. 600-601, especially Fig. 19.4
2. p. 601-602, Fig. 19.5
3. p. 602-603, Fig. 19.5
4. p. 603
5. p. 603
6. p. 603-604, Fig. 19.4
7. p. 603-604
8. p. 604-605
9. p. 604-605
10. p. 605
11. p. 608-612, Fig. 19.14, beach cross-section in the art spread, p. 610
12. p. 609, Diagram of longshore transport in the art spread, p. 611
13. p. 612-613
14. p. 613-614
15. p. 614-615
16. p. 606-607
17. p. 607
18. p. 615-617
19. p. 617-618
20. p. 620-622
21. p. 617-618

Completion Questions

1. low / storm
2. flood / ebb
3. spring / neap
4. spring / neap
5. wind
6. crest / trough / height
7. circular
8. wave base / one half
9. seas
10. swells
11. rogue wave
12. fetch
13. breakers
14. decrease / decrease / increase
15. parallel
16. Rip / sea
17. beach
18. storms / berms
19. foreshore
20. drift
21. groins
22. spits
23. baymouth
24. island / mainland
25. parallel / lagoon
26. wave-cut
27. seastacks
28. emergent
29. submergent
30. terraces

Multiple Choice

1. b	6. a	11. c	16. b	21. d
2. a	7. d	12. b	17. a	22. d
3. d	8. b	13. a	18. b	23. d
4. c	9. b	14. d	19. c	24. c
5. d	10. a	15. b	20. c	25. a

True or False

1. F	6. F	11. F	16. F	21. F
2. F	7. T	12. F	17. F	22. T
3. F	8. T	13. T	18. F	23. T
4. T	9. T	14. F	19. T	24. F
5. F	10. T	15. T	20. T	25. T

Drawings and Figures

1. Fig. 19.5
2. Figs. 19.15a, 19.16
3. (a) barrier island, (b) wave-cut platform

Chapter 20

Chapter Concept Questions

1. p. 632
2. p. 633
3. p. 633-634
4. p. 637
5. p. 639

Completion Questions

1. ash/ fertile
2. concrete/ pumice
3. increased/ deforestation
4. 40
5. tons
6. energy/ impact
7. where/ how
8. soil/ water
9. 10X
10. 42,000

Multiple Choice

1. a	3. c	5. c	7. a	9. c
2. b	4. d	6. b	8. b	10. d

True or False

1. T	3. T	5. F	7. T	9. F
2. F	4. F	6. T	8. T	10. T

Drawings and Figures

1. Fig. 20.9b, p. 638
2. Fig. 20.14, p. 643